DEPARTMENT OF NATIONAL RESOURCES
AUSTRALIAN WATER RESOURCES COUNCIL

RESEARCH PROJECT No. 72/47

MULTI-OBJECTIVE PLANNING MANAGEMENT FOR WATER RESOURCES – REVIEW OF RECENT DEVELOPMENTS IN THE U.S.A.

by
W. T. O'Brien,
Centre for Environmental Studies
University of Melbourne

AUSTRALIAN WATER RESOURCES COUNCIL
TECHNICAL PAPER No. 29

AUSTRALIAN GOVERNMENT PUBLISHING SERVICE
CANBERRA 1977

Published for the Department of National Resources on behalf of the Australian Water Resources Council by the Australian Government Publishing Service.

The AWRC Technical Paper series presents information resulting from research projects supported by the Water Research Fund of the Australian Water Resources Council. The report is presented in accordance with general policy of providing for the wide dissemination of research findings arising out of the Council's research program. Publication of the report in this series does not signify that the contents necessarily reflect the views and policies of the Australian Water Resources Council or the Commonwealth or State Governments. Details of other publications in the series are given on the back cover.

ISBN 0 642 03196 7

Printed by Canberra Reprographic Printers, Fyshwick A.C.T.

PREFACE

This report reviews recent developments in the application of the U.S. Principles and Standards for planning water and related land resources. It follows an earlier AWRC report (O'Brien et al, 1977) which examined the Principles and Standards for its suitability as a possible starting point in the development of a set of planning guidelines relevant to Australian conditions.

The report has two major purposes. First, it reviews several studies on planning policy which have been undertaken by the U.S. Water Resources Council since publication of the Principles and Standards in September 1973; these studies have investigated various policy options for the objectives, discount rate and cost-sharing arrangements used in the Principles and Standards. Council studies on possible extension of the program coverage of the Principles and Standards and the development of planning policies to integrate the Environmental Impact Statement requirements of the National Environmental Policy Act of 1969 into the planning requirements of the Principles and Standards are also outlined. The second part of the report reviews the development and application of procedures to implement the Principles and Standards by the U.S. Army Corps of Engineers, the largest of the U.S. Federal water agencies.

Preparation of this report has involved discussions with many staff members of U.S. water agencies. In particular, the assistance of the following is gratefully acknowledged:

Mr. Garry Cobb (Acting Director), Mr. Frank Thomas and Mr. Bob Waters, U.S. Water Resources Council, Washington, D.C.

Mr. William Donovan (Chief), Mr. Ed. Cohon, Mr. Don Roeseje and Mr. Alexander Otto, Plan Formulation and Evaluation Branch, U.S. Army Corps of Engineers, Washington, D.C.

Mr. Dick Porter, Planning Division, U.S. Bureau of Reclamation, Washington, D.C.

Mr. Alan Atkins, Centre for Environmental Studies, gave valuable assistance in reviewing the cost-sharing and discount policy studies undertaken by the U.S. Water Resources Council. The care and attention given to report typing by Mrs. Sandra Guiney is also gratefully acknowledged.

The report was made possible by a grant from the Commonwealth Government in support of a research program approved by the Australian Water Resources Council. The author is grateful to the officers of the Water Branch of the Department of National Resources, which provides the AWRC Secretariat, for their valuable assistance during preparation of the report.

CONTENTS

TABLES

1. INTRODUCTION

1.1. Setting to the Report

To provide a setting for the report, the following sections briefly outline the organisational basis of U.S. water management, the historical development of the U.S. Water Resources Council's Principles and Standards for planning water and related land resources and the scope of the present coverage of the Principles and Standards.

1.1.1. Organisation of U.S. Water Management

Planning, development and management of water and land resources in the United States take place under a federal system of government. The Federal Government was established under a concept of limited and constitutionally delegated powers; the States, by contrast, retain all residual powers of government not constitutionally prohibited or delegated to the Federal Government. Over the years, these delegated powers have been increased by constitutional amendment, interpreted broadly by the Supreme Court, and implemented by Congressional action to provide a significant body of law and authority for Federal Government activities, including water resource development, preservation of wild and scenic rivers and regulation of water pollution.

This evolution of Federal jurisdiction has changed the division of responsibilities between Federal and State Governments. Today, State Governments are concerned primarily with private property rights in water and land and related regulatory and property taxation systems; the States may, however, construct large engineering works and engage in major water management schemes, and a few States - most notably the State of California - have done so. Municipalities and other forms of local government continue to have major responsibility for domestic water supply and distribution systems and for waste collection and treatment. Where inter-related problems are involved, various organisational arrangements are being utilised to bring the concerned local, State and Federal agencies together. Large investments in water resources developments have also been made by private interests for industrial processing, energy production, transportation, and other purposes.

Within the Federal Government there are seven departments with 18 agencies and seven independent agencies that have water programs; these agencies have different goals, clientele, and organisational arrangements. The 25 individual agency water programs are funded through a total of 70 appropriations accounts. Eighteen of the 25 agencies are involved in preparing water plans which range from specific projects to comprehensive area-wide plans. At the Federal level of government the long-standing organisational issue has been whether to consolidate related water resource functions into a Department of Natural Resources or to co-ordinate policy and procedures for the conduct of such functions through a board or council composed of representatives of the functional agencies. In 1965 Congress opted for the latter alternative by passage of the Water Resources Planning Act (Public Law 89-80). Among other things this Act established the U.S. Water Resources Council '... to encourage the conservation, development and utilisation of water and related land resources on a comprehensive and co-ordinated basis...' by the Federal, State and local governments and private enterprise.

The members of the U.S. Water Resources Council are the Secretaries of the Departments of Agriculture, Army, Commerce, Interior, Housing and Urban Development, and Transportation, the Administrator of the Environmental Protection Agency and the Chairman of the Federal Power Commission. The Administrator of the Federal Energy Administration is an Associate Member. The Director of the Office of Management and Budget, the Attorney General, the Chairman of the Council of Environmental Quality, the Chairman of River Basin Commissions, and the Chairman of the Tennessee Valley Authority are Observers. Other departments and agencies are invited to Council meetings when matters pertaining to their responsibilities are under consideration.

1.1.2 Development of the Principles and Standards

The U.S. Principles and Standards for planning water and related land resources has evolved in an attempt to find a balance between economic and environmental considerations in water planning.

Congressional intent in the Flood Control Act of 1936 that benefits of Federal projects should exceed costs led to planning procedures for evaluating the benefits and costs of proposed projects. These procedures were based on a national economic efficiency analysis and were first published as 'Proposed Practices for Economic Analysis of River Basin Projects' in 1950. Budget Bureau Circular No. A-47 was issued at the end of 1952, informing the agencies of considerations which would be used in evaluation of projects and requiring uniform data that would permit comparisons among projects.

In October 1961, the President requested the Secretaries of the Interior, Agriculture, Army, and Health, Education and Welfare to review existing evaluation standards and to recommend improvements. The resulting report, 'Policies, Standards, and Procedures in the Formulation, Evaluation, and Review of Plans for Use and Development of Water and Related Land Resources', was approved by the President on May 15, 1962, and published as Senate Document No. 97, 87th Congress, Second Session. This document replaced Budget Bureau Circular No. A-47.

Senate Document 97 for the first time recognised development, preservation, and well-being of people as co-equal objectives. However, plans were to be formulated initially on the basis of economic benefits and costs and then adjusted to take account of intangibles such as preservation (e.g., wild and scenic rivers) and well-being of people (e.g., death and hardship from floods). In actual plan formulations during subsequent years preservation and well-being were not given co-equal consideration with development.

Since the early 1960 s, Congress has enacted many laws that have given new and more definitive directions for the consideration of environmental objectives in planning for water and land resources. Of particular relevance have been

- The Wilderness Act of 1964 which precluded further development of water resources in authorised wilderness areas of National Forests.

- The Land and Water Conservation Fund Act of 1965 which provides funds for Federal acquisition of lands and water for recreation purposes in National Forests, National Parks and National Wildlife Refuges and for matching grants to State and local governments for acquisition of recreation lands and waters.

- The Federal Water Project Recreation Act of 1965 which provides for full consideration of opportunities for recreation and fish and wildlife enhancement in Federal projects under specified cost allocation and cost-sharing provisions.

- The Water Quality Act of 1965 and the Clean Waters Restoration Act of 1966 which provided for establishing water quality standards and substantial Federal funds in aid of municipal sewage treatment.

- The Wild and Scenic Rivers Act of 1968 which provides that in planning for the use and development of water and related land resources consideration shall be given to potential wild, scenic and recreational river areas in river basin and project plan reports, and comparisons shall be made with development alternatives which would be precluded by preserving these areas. The Act also authorised establishment of eight national wild or scenic rivers and authorised studies of twenty-seven other rivers or parts thereof. Subsequently, both Federal and State governments have added areas to the national and State wild and scenic river systems and additional studies have been authorised.

- The National Environmental Policy Act of 1969 which directs Federal agencies in the decision-making process to give appropriate consideration to environmental amenities and values. The Environmental Impact Statements required by this Act have become a major source of information in public consideration of environmental questions. Through judicial actions in response to suits brought by environmentalists, the construction of many Federal water resource development projects has been modified or abandoned.

- The Water Pollution Control Amendments Act of 1972 which greatly strengthened governmental regulation of water polluting activities and placed substantial emphasis on planning problems and complementarities falling within the scope of a broadened Federal multi-objective planning emphasis.

After reviewing experience gained from implementation of Senate Document 97 and taking account of the many Congressional directives and other expressions of concern for both environmental and economic considerations, the Water Resources Council, in early 1969, undertook the development of a set of principles and standards for planning water and land resources. These were approved by the President in September 1973, and replaced Senate Document 97.

The Council's Principles and Standards represent the first attempt ever undertaken to formalise the application of multi-objective planning methodology* to the management of water resource systems. They provide that plans are to be formulated in terms of their beneficial or adverse effects on the objectives of national economic development and enhancement of environmental quality. Progress toward the national economic development objective depends on increasing the value of the nation's output of goods and services and improving national economic efficiency. Progress toward the environmental quality objective depends on the management, conservation, preservation, creation, restoration, or improvement of the quality of certain natural and cultural resources and ecological systems. Action to produce economic benefits may produce complementary environmental benefits but often there is conflict and benefits to one objective must be given up to achieve more of the other. To aid decision-makers in selecting among alternative courses of action, there must be a complete display of relevant beneficial and adverse effects on these two objectives for each alternative plan.

1.1.3. Program Coverage of the Principles and Standards

When the Principles and Standards were promulgated in 1973, they were restricted in coverage to direct Federal water programs, i.e., programs which were undertaken by Federal agencies such as the Corps of Engineers, Bureau of Reclamation and Soil Conservation Service. The Principles and Standards were not made binding on Federal grant and loan programs i.e. programs for which non-Federal interests receive *monies* from the Federal Government in the form of grants and loans, respectively.

During the past 5 years, the changing picture of U.S. Federal water program investments has significantly altered the percentage of the total Federal water program covered by the Principles and Standards.

In 1972, Federal investments for water totalled $5.2 billion; by 1976 this figure had grown to $10.2 billion. The major factor in this doubling has been the rapid increase in investment by the Environmental Protection Agency (EPA) in water quality improvement (from less than $1 billion in 1972 to more than $5.4 billion in 1976). During this time the water programs of the Departments of the Army, Interior and Agriculture have either decreased or barely held their own against inflation. Because the EPA investments are primarily through grant and loan programs, neither of which fall within the scope of the Principles and Standards, *only 41 percent of the Federal water investments are presently covered by the Principles and Standards.*

---

* For a full discussion of multi-objective planning methodology, see O'Brien et al (1977), Chapter 1.

This increased growth in Federal grant and loan programs, outside the planning framework of the Principles and Standards, has caused the Water Resources Council to adopt the following position on program coverage:

> '... *all* Federal and Federally assisted water and related resource programs must relate to a national policy framework. It was decided that ... a set of principles will be developed, and then consideration will be given to appropriate standards ...'
>
> *WRC 1976 p13 (emphasis added)*

As a first step in initiating this Council position, the Federal departments with water grant and loan programs (Interior, HUD, Agriculture, Commerce, WRC and EPA) have developed papers outlining the scope of such programs, problems associated with coverage under a common set of principles and recommendations for resolving such problems. Work has now progressed to the point where a single set of *Principles** is being developed to cover both direct Federal programs and Federally assisted (grant and loan) programs. However, in recognition of the inherent differences in legislative authorities, Congressional interest and decision making and management control processes which exist between grant and loan programs and direct Federal programs, different *Standards*** are being developed for the two types of programs.

1.2. Scope and Purpose of Report

The aim of this report, as stated in the Project Schedule, is to *outline and review recent developments in the application of the 1973 Principles and Standards in the U.S.A. and to note their possible implications for Australian application.* This report follows an earlier AWRC report (O'Brien *et al,* 1977) which reviewed the 1973 Principles and Standards for their suitability as a possible starting point in the development of a set of guidelines for water resources planning in Australia.

At the policy level, the most important review of the Principles and Standards has been that undertaken by the U.S. Water Resources Council in response to the requirements of Section 80(c) of the 1974 Water Resources Development Act (P.L.93-251). Under this section of the Act the President was directed to conduct a full and complete investigation and study of the Principles and Standards including, but not limited to, *planning objectives, discount rate* and *cost sharing* as these relate to Federally-financed water and related resource projects. The results of the Council's study were submitted to the President in December 1975 but as yet no recommendations on the study issues have been submitted to Congress. The Section 80(c) study report is reviewed in Section 2 of this report.

---

* The *Principles* provide the broad policy framework for planning activities and include the conceptual basis of planning.

** The *Standards* provide for uniformity and consistency in comparing, measuring and judging beneficial and adverse effects of alternative plans.

At the implementation level, planning procedures conforming to the Principles and Standards have now been approved for most agencies involved in the planning of Federally-financed water projects, including the U.S. Army Corps of Engineers, Bureau of Reclamation, Soil Conservation Service, Tennessee Valley Authority, National Parks Service and Bureau of Outdoor Recreation. The experience gained by the Corps of Engineers (the largest of these agencies), in implementing the Principles and Standards and in integrating the Environmental Impact Statement requirements of the National Environmental Policy Act of 1969 into the planning process, is reviewed in Section 3.

Conclusions on U.S. experience in the application and further development of the Principles and Standards, together with recommendations on their possible implications for Australian applications are presented in Section 4.

1.3. Future Developments

This present report should be regarded only as an interim statement of U.S. experience with the Principles and Standards - one that summarises the major policy and implementation developments during the period from September 1973 until May 1977. During this period, many issues have been raised and many options have been proposed, but much remains unresolved. The next twelve to eighteen months, however, may see not only the resolution of many of the present issues but also, under the Carter administration's strong thrust towards increased efficiency of use of both water and energy, the development of new directions in national water policy and corresponding revision of the Principles and Standards.

In his environmental message to Congress on May 23, 1977, President Carter ordered the Office of Management and Budget, the Council on Environmental Quality and the Water Resources Council to review current water resources policy and, after consultation with the Congress and the public, to report firm water policy recommendations within six months. The following day, U.S. Secretary of the Interior Andrus, in addressing the 1977 National Conference on Water at St. Louis, Missouri, defined nine points upon which a comprehensive national water policy should be based:*

(i) The planning and evaluation criteria in the Principles and Standards should be revised 'to reflect the President's policy initiatives and concepts' and the revised Principles and Standards should provide the framework for the Council's review of national water planning policy. Modifications to the Principles and Standards should include the following revisions:

---

* The full text of this important statement on U.S. water policy and planning is reproduced in Appendix A.

(a) Water conservation* should be added as an objective for consideration with the national economic development and environmental quality objectives.

(b) Equal standing should be given to plans which achieve environmental conservation and economic objectives through non-structural means and with explicit consideration of the alternative uses of water. Flood plain management and flood insurance should have high priority.

(ii) Projects that have been authorised for more than eight years and not yet funded or under construction should be de-authorised so that an up-to-date analysis of such projects can be undertaken before their re-authorisation.

(iii) Increased cost sharing by the States and other non-Federal entities is essential because the present system 'encourages waste where the price of water is below the actual cost'.

(iv) Co-operative efforts between the states and the Federal government must be undertaken to eliminate laws, rules and institutions which now hamper integrated water management.

(v) All Federal programs should be scrutinised to seek new methods of encouraging water conservation.

(vi) Quantification of 'reserved rights' for Indian reservations and other Federal lands is essential to rational planning for the future.

(vii) Water quality and waste water management planning need to be evaluated as part of conventional water resources planning: '... waste water reuse, artificial ground water recharge, minimum streamflows and safe drinking water are primary objectives - not just afterthoughts or appendages'.

(viii) Safety of water projects cannot be taken for granted - it must be established and proven in all cases.

(ix) The primary goal of water planning is conservation and more efficient use of water.

---

* In Andrus's statement *water conservation* is used in the same sense as energy conservation and means the promotion of greater efficiency in water use through implementation of *demand* management techniques, such as pricing policies, rather than through traditional supply management techniques, such as dam construction. This emphasis on demand management is consistent with the major emphasis on energy conservation adopted by President Carter in his National Energy Policy proposal to Congress in April 1977.

All the above issues will have some bearing on the future form and applicability of the Principles and Standards, either through modification of the planning objectives, plan formulation, assessment and evaluation procedures, program coverage or cost sharing arrangements. The basic multi-objective approach of the Principles and Standards, however, will almost certainly remain intact.

2. PLANNING AND COST SHARING POLICY OPTIONS

The Water Resources Council completed in December 1975 a study of the Principles and Standards with emphasis on the development of new options for the *planning objectives, discount rate* and *cost sharing*. This study was assigned to the Council by the President under the authority of Section 80(c) of the 1974 Water Resources Development Act (P.L.93-251). The technical information prepared by the Council study team was published as a twenty-volume document entitled *Planning and Cost Sharing Policy Options for Water and Related Land Programs* (WRC, 1975). At the present time (July 1977), the various planning and cost sharing options presented in the study report are still under review by the Office of Management and Budget. These options are summarised below.

2.1. Planning Objectives

Objectives serve two basic functions in water resources planning: (i) they guide plan formulation at the field or planning level, and (ii) they provide a basis for plan evaluation at the decision-making level. Specification of particular objectives for planning under the Principles and Standards has implications at both levels:

> Objectives in the Principles and Standards are defined in broad categorical terms. This permits the plans developed under the Principles and Standards to assume different characteristics reflective of the benefits and costs, consistent with the specified objectives, of alternative water resource management measures in each specific planning setting. The Principles and Standards provide that an array of alternative plans be developed in each planning setting. Each of the alternative plans is to represent the best way to serve a different objective or mix of objectives. *The more objectives there are, the more complex the trade-offs become. Consequently, more plans must be developed to display the trade-offs among the objectives as additional objectives are added.* In each study, a recommended plan which serves the combination of objectives which best reflects public priorities and preferences is to be selected from the array of alternative plans ...
>
> Once the planning objectives are specified, planners can develop (formulate) plans and appraise (evaluate) the merits of such plans in terms of the specified planning objectives. These plans will

include combinations of structural and non-structural measures for the management of water resources. Each measure will have different effects on the objectives. Thus, changing the objectives included in the multi-objective specifications will change:

.. The kinds of water resource management measures considered;
.. The selection of alternative measures;
.. The size or scale of measures; and
.. The siting of structural measures.

For decision-makers, that is, agency heads, officials of the Executive Office of the President and the Congressional Committees ... objectives provide standards against which to measure the value of water resource plans. A plan which is unattractive under one set of objectives may become desirable under a different set of objectives. It is important to note that a plan may be evaluated and selected or rejected on the basis of criteria other than those used to design (formulate) the plan. For this reason, the WRC Principles provide for a full display of the impacts of a plan in its system of four accounts. It should be recognised, however, that if key decision-makers consistently use different criteria (objectives) to evaluate and select plans, planners will tend to substitute the key decision-makers' criteria for the nominal criteria in the formulation of individual water resource projects.

WRC 1975a pp4, 5 (emphasis added)

The major options to the planning objectives used in the current Principles and Standards *(national economic development and environmental quality)* were presented in Parts 3A and 3B of the study team's publications (WRC, 1975a,b). The options discussed in these papers reflected two radically different philosophies towards multi-objective planning: Paper 3A retained the basic *multi-objective approach* used in the Principles and Standards; Paper 3B explored two alternative approaches, the *integrated objective approach* and the *constrained economic maximisation approach**. These options are discussed below.

2.1.1. Multi-objective Approach (Principles and Standards)

Using the existing multi-objective planning framework Paper 3A presented options which are combinations of the objectives presently included as objectives or implied in the accounts in the present Principles and Standards. It began from the premise that there is general consensus regarding the utility of estimating economic benefits and costs and that the *NED objective should be included in any specification of multi-objectives.* From this starting point

---

* The constrained economic maximisation approach (Section 2.1.3.) is the traditional planning approach of Australian and U.S. water agencies.

the paper argued that '...there are a series of decision points where choices can be made about environmental quality, regional development and social well-being'. The important statement was made that:

> 'None of the options presented in the paper imply any changes in the accounts as presented in the Principles and Standards or in the requirement to display all significant impacts of a plan regardless of how it is to be formulated'.
>
> *WRC 1975a p3*

The statement of National Economic Development (NED) and Environmental Quality (EQ) objectives in the Principles and Standards is as follows:

> The overall purpose of water and land resource planning is to promote the quality of life, by reflecting society's preferences for attainment of the objectives defined below:
>
> A. To enhance national economic development by increasing the value of the Nation's output of goods and services and improving economic efficiency.
>
> B. To enhance the quality of the environment by the management, conservation, preservation, creation, restoration, or improvement of the quality of certain natural and cultural resources and ecological systems.

On the assumption that '... a decision about one objective is independent of decisions about other objectives', the paper presented three non-mutually exclusive options for modifying the statement of objectives in the Principles and Standards. The first option was to expand the environmental objective as follows:

> B'. To enhance the quality of the environment by the management, conservation, preservation, creation, restoration, or improvement of the quality of the Nation's natural and cultural resources and ecological systems.

The second option was to add regional economic development (RED) as an objective:

> C. To increase the total and/or per capita income of certain selected regions beyond what is possible under the national economic development objective through additional transfers of economic activity.

The third option was to add the life, health and safety component of the existing social well-being account as an objective:

D. To enhance the social well-being of people by contributing to their security of life, health and safety.

As noted in the paper:

> *Each of the above statements of objectives relates to a specific category of impacts resulting from water resource plans. The concepts reflected in the proposed definitions of objectives have operational meaning and are mutually exclusive. The objectives as defined above can be understood and applied consistently in specific planning situations. Moreover, each objective reflects a category of impacts not reflected explicitly in any other objective.*
>
> *Ibid. pp 7, 8 (emphasis added)*

Option 1 - Expansion of the Environmental Quality Objective

The first option proposed in Paper 3A was the modification of the existing EQ objective by substituting the words 'The Nation's' for the word 'certain' in the definition of the objective. The definition of the EQ objective in this option would permit consideration of all environmental gains for the Nation as a whole in plan formulation, but it would exclude consideration of transfers of EQ effects which did not result in a net improvement for the Nation as a whole.

It was argued in the paper that adoption of this option would eliminate the confusion that has arisen in interpretation of the word 'certain' in the existing Principles and Standards. By redefining the environmental objective to encompas *all* net improvements in environmental quality within the Nation, it would avoid the situation under the existing definition of the EQ objective where, for example, environmental goods which were considered valuable only by the people in a particular study area could be excluded from consideration in plan formulation on the grounds that such goods were duplicated elsewhere in the Nation.

Option 2 - Addition of a Regional Development Objective

The second option proposed in Paper 3A was the inclusion of regional economic development as an objective. As noted earlier, the regional economic development objective was defined as:

> 'To increase the total and/or per capita income of certain selected regions beyond what is possible under the national economic development objective through additional transfers of economic activity'.

As noted in the paper, this definition is more limited in its scope than the components of the regional development account in the Principles and Standards. In addition to regional income, the regional development account in the Standards includes the following components:

- regional employment;
- population distribution;
- regional economic base and stability; and
- environmental conditions of special regional concern.

It was argued in the paper that regional employment, population distribution and regional economic base stability are all closely related to total and/or per capita regional income and because it is easier to measure the latter, plans should be formulated in terms of regional income. It was also argued that where regional employment, population distribution and economic stability are not related to income, their relationship to water resource projects is obscure, and that environmental conditions of special regional concern which constitute net gains to the Nation are reflected in the option for the restated environmental quality objective (Option 1).

The major implication in choosing regional economic development as a planning objective would be the achievement of a different distribution of income among regions at the expense of a smaller national income. As noted by the study:

> 'Planning for a regional economic development objective also means that environmental quality as well as net national economic benefits may be foregone in the interest of promoting the economic development of a particular region. A plan formulated to serve the regional economic development objective need not necessarily have net national economic benefits or be desirable from an environmental quality perspective. It is possible, however, to constrain plan formulation under the regional economic development objective by requiring, for example, that any plan formulated to that objective must still have national economic benefits at least as great as national economic costs'.
>
> *Ibid. pp 13, 14.*

It was noted that adoption of the regional economic development objective would require the selection of regions to which income transfers are desirable. Some alternative approaches to the selection of target regions based on the use of economic criteria were presented in Appendix B of the report.

<u>Option 3 - Addition of a Life, Health and Safety Option</u>

The third option proposed in Paper 3A was the adoption of the life, health and safety component of the social well-being account in the Principles and Standards as an objective. As noted earlier, this objective would be defined as:

> 'To enhance the social well-being of people by contributing to their security of life, health and safety'.

It was noted that the original social well-being objective, as described by the Task Force which developed the first version of the Principles and Standards in 1969, was clearly designed to be a catch-all objective that would include effects of plans not reflected in the NED, EQ or RED objectives. Social well-being (SWB), which was retained as an account in the Standards, includes the following components:

- income distribution;
- security of life, health and safety;
- educational, cultural and recreational opportunities;
- effects on emergency preparedness; and
- other.

The paper argued that:

> 'An objective derived from this broad consideration would, therefore, be a loosely worded, open-ended, generalised statement. Such an approach was not taken in this paper for three reasons. First, such a definition would lack generally agreed upon operational meaning. That is, it would not be possible to define, in a given situation, what constituted a significant change in social well-being and to formulate plans accordingly. Second, although water resource projects do have impacts on considerations such as the distribution of income among classes (i.e., rich versus poor), this is the result primarily of cost sharing policies as opposed to the actual design of the project itself. Third, some of the components of the SWB account, such as educational, cultural and recreational opportunities, are already provided for in the EQ and NED objectives. This paper, therefore, provides an option for adding only life, health and safety to the array of objectives. Regardless of whether the life, health and safety objective is adopted, all effects on components now included in the existing social well-being account will continue to be reported in planning reports when they are significant'.
>
> *Ibid. pp 15, 16*

The life, health and safety objective could influence plan formulation and evaluation by changing the type,scale or location of measures included in a water resources plan. As an example, the design of flood protection measures based on economic efficiency might result in protection against floods with a recurrence interval of 70 years. Plan formulation for social well-being, however, could result in protection against larger floods with a less frequent recurrence interval. The additional protection, although not justified in terms of national economic development benefits, would be included to enhance the security of life, health and safety of people living in the floodplain.

Assessment and Evaluation of Options

Tables I and II summarise the assessment of the study team regarding the expected effects of the various options for planning objectives at the *project* and *program** levels, respectively.

At both levels, the effects of restating the environmental objective are the most uncertain. This reflects the absence of WRC and agency policy statements defining the scope of the current environmental quality objective.

As shown in Tables I and II, the addition of regional economic development as an objective would have important effects on both project and program formulation. In regions selected for regional development, the scale of projects formulated under the RED objective would be increased beyond that justified by economic or environmental considerations alone. It would provide a basis for authorising projects which might not otherwise be justified. The effects of such actions at the project level would then mean that, at the program level, fewer projects could be built for the Nation as a whole (for a given level of national expenditure on water resources development).

Of the programs now covered by the Principles and Standards**, life, health and safety considerations appear to be relevant to flood control, drainage and water supply studies. However, the study team acknowledged that adopting these social well-being considerations as an objective would have little effect on project formulation for flood control and water supply because current agency procedures already take health and safety factors into account in these cases as a public policy. If however the Principles and Standards are extended to cover programs dealing with water quality (as appears almost certain to happen) human health would be a fundamental element in program justification and life, health and safety could be the dominant objectives in planning for such programs.

Table III summarises the major issues which the study team regarded as important in choosing objectives for water resource planning. The upper part of the table indicates reasons which might lead decision-makers to choose a particular option. The lower part of the table displays the implications of choosing that option. As noted previously, preference sets II, III and IV are not mutually exclusive, although, as noted in Paper 3A:

---

* A program is defined as a given multi-year budget allocation for water resource functions of all kinds, at either the national or regional level.

** Examples of programs not currently covered by the Principles and Standards include virtually all Federal grant and loan programs for water resource management and development. This includes the EPA construction grant program, Department of Commerce Economic Development Administration grants and several grant and loan programs administered by the Department of Agriculture. Collectively, grant and loan programs comprise the major part (about 60%) of Federal water resource expenditure.

TABLE I

PROJECT IMPACTS OF ADDITIONAL OBJECTIVES FOR THE FORMULATION AND EVALUATION OF ALTERNATIVE WATER RESOURCE PLANS

(TABLE III-I, WRC, 1975a)

Category of Project Impact	Restatement of Environmental Quality Objective	Addition of Regional Economic Development Objective	Addition of Life, Health and Safety Objective
Types of Measures (Structural and Nonstructural)	Leads to consideration of additional measures to enhance the quality of the Nation's natural and cultural resources and ecological systems.	No significant impact	No significant impact
Selection of Alternative Measures	Broader range of EQ effects considered in plan formulation and evaluation.	Adds consideration of effects on regional income to evaluation criteria: emphasises regional economic development at expense of EQ and NED in formulation of RED plan.	Adds explicit life, health and safety considerations in formulation of alternative plans. Provides added justification for plans for drainage and urban flood control.
Scale of Measures	Tends to reduce scale of plans with adverse impacts on the Nation's natural and cultural resources and ecological systems.	Increases scale of measures and may impact on mix of plan outputs.	Increases scale of some drainage and flood control plans beyond size formulated according to NED or EQ objectives.
Siting of Structural Measures	Results in broader concern for all net national EQ impacts.	Unknown	Results in greater concern for life, health and safety in evaluation of alternative sites.

TABLE II

PROGRAM IMPACTS OF ADDITIONAL OBJECTIVES FOR THE FORMULATION AND EVALUATION OF ALTERNATIVE WATER RESOURCE PLANS

TABLE III-2, WRC, 1975a

Category of Program Impact	Restatement of Environmental Quality Objective	Addition of Regional Economic Development Objective	Addition of Life, Health and Safety Objective
Project Mix	Broadens emphasis on EQ	For a given level of expenditure, larger but fewer projects could be built.	For a given level of expenditure, fewer projects could be built because individual flood control or drainage projects would be more expensive.
Geographic Distribution of Projects	Unknown	Areas for which regional development was an objective would receive a larger share of a given national level of expenditure. For a given level of regional expenditure, a region would receive larger but fewer projects.	For a given level of expenditure on flood control, some areas would receive a higher level of protection than could be justified under the NED or EQ objectives. Other areas, however, would be denied any protection. Thus, benefits would become more concentrated geographically.

TABLE III
PREFERENCE SETS FOR OPTIONS FOR PLANNING OBJECTIVES
TABLE III-3, WRC, 1975a

PREFERENCE SET I	PREFERENCE SET II	PREFERENCE SET III	PREFERENCE SET IV
In order to Reaffirm the Current Statement of Objectives in the Principles and Standards, Decision-Makers Should Want to:	In Order to Restate the EQ Objective, Decision-Makers Should Want to:	In Order to Add Regional Economic Development to the Objectives Decision-Makers Should Want to:	In order to Add Life,Health and Safety to the Objectives, Decision-Makers Should Want to:
1. Restrict the objectives to be considered in plan formulation to NED and EQ; and 2. Retain a definition of EQ which limits the range of EQ effects to be considered in plan formulation.	1. Encompass, within the EQ objective, all net improvements to environmental quality for the Nation as a whole.	1. Favour economic development in certain selected regions; and 2. Achieve regional economic development through projects different from those formulated under the national economic development objective 1/	1. Separate explicity life, health and safety from economic development and environmental quality, that is, to provide for life, health and safety aspects of social well-being in a way that is not provided for under the economic and environmental objectives.
And Be Willing to:	And Be Willing to:	And Be Willing to:	And Be Willing to:
1. Define in the Standards and in the procedures the range of EQ effects to be considered in plan formulation; and 2. Forgo the contributions of additional objectives to the plan formulation process.	1. Accept a change in policy regarding environmental quality before the results of planning under the current WRC approach to EQ can be evaluated fully.	1. Accept reductions in the net national economic benefits and the environmental benefits of water resource programs to the extent that RED is reflected in recommended plans; 2. Specify regions for which regional economic development would be an objective; 3. Accept interregional transfers of economic activity beyond that associated with national economic development; and 4. Add to the number of plans which must be developed to display the trade-offs among the objectives.	1. Accept additional projects which contribute to human life, health and safety as well as larger projects than are justified under economic and environmental criteria to the extent that life, health and safety considerations are incorporated into recommended plans; and 2. Add to the number of plans which must be developed to display the trade-offs among the objectives.

1/ Both conditions must be met to make regional economic development a planning objective for project formulation. Alternatively, regional economic development could be included among the criteria used in allocating water resource expenditures among regions.

'It must be kept in mind that while each possible objective brings an added dimension to the plan formulation process, each objective places an added burden on the planning process by making it more complex and expensive'.

Ibid. p.22

Of the various options examined, the study team felt that the most important and far-reaching decision probably would be to adopt regional economic development as an objective for project planning. The study noted that a decision to include regional economic development as an objective for project formulation in selected regions would imply a corresponding use in program formulation to be effective. On the other hand, it was pointed out that a decision not to include regional economic development as an objective for project formulation does not preclude its use in program formulation - under the program approach projects formulated under the Principles and Standards would continue to be formulated to the same national objectives (NED, EQ and possibly the life, health and safety objective) but projects of all kinds (water and other government agencies) would be given *funding* priorities if they fell within the regions selected for regional economic development.

2.1.2. Integrated Objective Approach

Paper 3B (WRC, 1975 b) presented an alternative approach for plan formulation to that used in both the Principles and Standards and Paper 3A (WRC, 1975 a). This approach, termed the *integrated objective approach,* incorporates the economic, environmental and social effects (both beneficial and adverse) of a plan into one account and summarises these effects in a single dollar measure of 'net social worth' defined as being:

'... the algebraic sum of the total value of output of economic, social and environmental goods minus all costs incurred in the provision of those goods'.

WRC 1975b p7

The integrated objective approach is similar to benefit-cost methodology in that it leads to a single number; its additional (and controversial) feature is that it requires the systematic categorisation, quantification and pricing of all goods and services, *including those not priced in the market place* (such as environmental quality and life, health and safety effects). As noted in Paper 3B:

'The validity of the integrated objective approach depends upon the ability of society not only to make explicit collective social value judgements but to express them in comparable units, i.e., dollars. This requires both technical and institutional innovation'.

Ibid. p8

and:

> 'In a similar manner (to the market economy), there are government policies and priorities, public interest groups, and general public concerns constantly interacting to express society's evaluation of social and environmental goods. The integrated objective approach postulates that, through analysis of these public and governmental expressions, it is possible to assign a price to non-market goods. Because the forces of the market place are obscured or absent from the evaluation of social and environmental goods, there will be a greater range of possibilities and much less certainty associated with a price for such goods. Public inputs will need to be systematically considered, and decision-makers will have a new and important responsibility. *The integrated objective concept focuses debate on trade-offs at the program as opposed to the individual project level.*'
>
> *Ibid. p9 (emphases added)*

The key to the utility of the integrated objective approach was thus clearly recognised to depend upon the ability to assign prices to non-market effects at the *program* level of planning. To guide this process, Paper 3B suggested that appropriate prices for non-market effects of plans must satisfy three *criteria* and that these prices must be set by some *institution* (to be established). The specified *criteria* were:

'1. Consistency and Agreement - There must be widespread consensus about the price assigned to a non-market value of a good. This consensus should be consistent among the public, planners at the field level, and decision-makers evaluating a plan. Proper consensus about assigned prices should imply a collective willingness to pay.

2. Uniqueness - The price of a good should be a direct function of the uniqueness or quality of the particular good. This implies that goods with a widely different quality or availability are in fact different goods and should each have a different price. To satisfy this criterion, much work is needed to develop meaningful categories of effects. Prices cannot be developed until a quantifiable taxonomy of effects has been developed.

3. Constancy Through Time - It should be expected that the price assigned different non-market goods remains fairly constant relative to the price of market goods. The relative price of goods and services may change to reflect the changing values or priorities of society. However, care must be taken to avoid quick price changes in response to

> fads or short-term crises. To maintain a given set of relative prices, prices of non-market goods should have to be inflated over time to reflect inflation of market prices.'
>
> *Ibid. pp10, 11*

The important elements to be considered in establishing an appropriate institution to determine the price of non-market goods were specified as:

> '... first, its ability to implement the above pricing criteria; second, its capacity to reflect public attitudes and preferences. The prices assigned will then be explicit, consistent and reflect public consensus.'
>
> *Ibid. p11*

It was suggested in the paper that one possible way for assigning non-market prices would be to establish a *Federal Valuation Board*, having the following *modus operandi:*

> '... the Federal Valuation Board would operate in areas where market forces are not effective or cannot be simulated.
>
> In order to reflect regional differences, the Federal Valuation Board would assign prices for non-market goods in each of the 21 WRC regions. The regional prices would be adjusted and refined by a process of public participation at the local and regional level. This is consistent with the call at the National Conference on Water for national policies which are sensitive to regional and local considerations. Before being considered final, the prices would be submitted by the Board to Congress for its consideration. Congressional response could take various forms. For example, the prices could be considered approved if no contrary action is taken in, say, 45 days.
>
> It is important that the prices assigned to non-market goods be determined before plans are formulated under the integrated objective approach. This is necessary to avoid the undue influence of a particular project and to give planners the requisite information for plan formulation. In this connection, it should be recognised that the Federal Valuation Board would not have responsibility to review individual projects. It would be concerned with individual projects only to the extent that implementation of projects changes the inventory ... and, hence the price ... of non-market goods.
>
> *Ibid. p12*

Given the prices assigned to non-market goods and services by the proposed Federal Valuation Board, an objective function for project planning could then be established. An iterative process of plan formulation would then be used to search for the plan which yielded the greatest value of this 'net social worth' function. The sensitivity of the preferred plan would then be tested by exploring the effect of small variations in the prices of the non-market goods.

2.1.3. Constrained Economic Maximisation Approach

Perhaps the most significant effect of the multi-objective approach of the Principles and Standards has arisen from the requirement that planners explicitly develop a range of plans responding to different mixes of the two national objectives of national economic development and environmental quality. This is a major departure from previous practice. In the United States, as in Australia, past emphasis has been on development of a 'best' plan which was oriented towards either maximising net economic benefits or maximising development of a water resource to the extent possible, within the constraint of having to achieve positive net benefits (a benefit-cost ratio greater than unity). More recently, agency planning has become increasingly sensitive to environmental concerns but, in the United States, prior to the promulgation of the Principles and Standards in 1973, these concerns were treated as constraints* rather than as positive goals towards which plans should be formulated. This approach has been termed in Paper 3B (WRC, 1975b) the *constrained NED (national economic development) maximisation approach.*

As noted in Paper 3B:

> 'A constrained NED approach would involve project planning to maximise the national economic development objective subject to designated enhancement or preservation level constraints for other objectives. These other objectives could include, depending upon which option for planning objectives is selected, environmental quality, regional economic development and life, health and safety'.
>
> *Ibid p6*

The report noted, however, that the *constrained NED maximisation approach* has one very important limitation:

> 'While this approach is conceptually simple to implement, it can lead to socially *optimum* plans only where the constraints established for the objectives other than NED can be set at the 'appropriate' level. *There is however, no accepted method for setting these constraints*'.
>
> *Ibid p6 (emphasis added)*

As an alternative to the multi-objective approach adopted in the Principles and Standards, the Section 80(c) study strongly supported the integrated objective approach (discussed in the preceding section) in preference to the constrained NED maximisation approach.

---

* In the same way as the benefit-cost ratio was treated as an economic constraint.

2.2. Discount Rate

Paper 4 (WRC, 1975c) presented three policy options for the discount (interest) rate used in the formulation and evaluation of U.S. Federal water and related land resource projects covered by the Principles and Standards. This paper is an important statement on the role of the discount rate in water planning and on the effects of the major discount rate options on project formulation and evaluation, including the effect of each option on the trade-off between economic and environmental objectives; it is included in full as Appendix B of this report. Some general comments on the role of the discount rate in planning and a brief summary of the options and their expected effects is presented below.

General Comments

In discussing the role of the discount rate in planning, the study team made three general comments which are particularly noteworthy:

1. The discount rate should be a 'real' rate and not a 'financial' rate, i.e. the economic benefits of goods and services received are compared to the economic costs in terms of goods and services required to produce the benefits and this comparison is made in constant dollars, e.g. 1975 dollars. Because economic evaluation is concerned with real-physical-resources rather than with financial flows, the effects of inflation are excluded from the estimation of project benefits and costs and observed market rates of interest are not directly applicable to water resources projects. This is the approach adopted in the Principles and Standards.

2. The discount rate should be a 'risk-free' rate, and risk and uncertainty should be dealt with explicitly in the estimation of project benefits and costs. This is the approach adopted in the Principles and Standards.

3. It is illogical to discount economic benefits and costs without discounting environmental benefits and costs, as is done in the Principles and Standards:*

   > '... given the function of the interest rate, it would be logical to apply it to both objectives provided values can be assigned to effects associated with each objective'.
   >
   > *Ibid. p6*

---

* For a contrary view, see Section 3.2. of this report.

The study team noted that the present approach used in the Principles and Standards affects the opportunity cost of obtaining environmental (and other non-market) benefits in terms of net economic benefits foregone.

Options

The paper examined the effects of three levels of the discount rate on the formulation and evaluation of water resource projects within the programs covered by the Principles and Standards: these rates '... can be related to but do not uniquely follow from the concepts of social time preference, opportunity cost and the cost of Federal borrowing' (p17). The rates assessed were three, six and ten percent.

The three percent rate corresponds approximately to the rate used by U.S. water agencies up until 1969. This rate is favoured by those who believe that a low interest rate is appropriate for water resource projects because the private time preferences which affect market rates of interest do not reflect sufficiently the collective wishes of society to provide for future generations (the 'social rate of time preference').

The six percent rate approximates the prescribed WRC rates for 1975 and 1976 which were determined by means of the formula provided for in Section 80(a) of the Water Resources Development Act of 1974 (P.L.93-251). This rate is related to the nominal cost, as opposed to the real cost, of long-term Federal borrowing.

The ten percent rate approximates the level desired by many who believe that the discount rate should represent the rate of return on investment in the private sector i.e., the return on water resource investments should be as great as the return from opportunities in the private sector which are foregone by such investments (the 'opportunity cost' of money).

All options provided for a fixed rate of interest to be used for a five-year period rather than one which changed annually. It was argued that not only would this action simplify the planning process but it would also strengthen the role of the interest rate in the design of water resource projects.

Assessment

The implications of the various discount rate options on project and program planning were summarised in Paper 4 as follows:

> In summary, a change in the interest rate affects the nature of the water resource program covered by the WRC Principles and Standards. This change is felt most strongly in projects oriented toward the NED objective. A change in the interest rate causes different kinds and different numbers of NED-oriented projects to emerge from the planning process. These differences include different outputs, different measures to

provide the outputs, different project sizes, as well as different patterns of costs and benefits over time.

Raising the discount rate favours smaller NED-oriented projects with high initial benefits (e.g., current water supply versus future water supply, reduction of flood damages to existing development versus future development). Moreover, a high discount rate favours producing outputs with measures with high operation and maintenance costs and low capital or investment costs. Conversely, lowering the discount rate favours larger NED-oriented projects whose benefits may not be realised until the future (e.g., development of a new waterway which would require several years for traffic to develop). Moreover, a lower discount rate favours measures with high initial investment costs and low operation and maintenance costs.

Although these conclusions focus on NED-oriented projects, a change in interest rate policy can also be expected to impact on projects formulated for the EQ objective or for a mix of NED and EQ objectives. For example, an increase in the discount rate should lead to the substitution of O&M costs for capital costs in all kinds of projects, EQ-oriented as well as NED-oriented. Paradoxically, however, an increase in the discount rate may result in larger EQ-oriented projects as well as a greater tendency to recommend these projects in some cases because the Principles and Standards do not require discounting of EQ benefits. This happens because a rise in the interest rate will often reduce the costs of obtaining EQ enhancements in terms of NED benefits foregone.

If the construction budget for programs covered by the WRC Principles and Standards expands or contracts in accordance with the number of justified projects, a set of conclusions can be drawn. As the interest rate rises, the total investment in projects oriented toward NED would decline, and the proportion oriented towards EQ would rise. There would be fewer projects as well as smaller projects oriented toward NED. Moreover, projects of all kinds tend to have higher OM&R costs. In this case, the size of the investment program would likely be constrained by the interest rate.

If the constraint on the water resource program is, instead, the budget level (that is, there are more justified projects at 'any' interest rate than can be funded), another set of conclusions can be drawn. The use of a high interest rate would permit more NED-oriented projects to be constructed because NED-oriented projects would tend to be smaller, and each dollar of investment expenditure would tend to place larger OM&R costs on future generations (which also enjoy the benefits). Conversely, the use of a low interest rate would permit fewer projects to be constructed because NED-oriented projects would tend to be larger. Moreover, each dollar of investment expenditure would place smaller OM&R costs on future generations.

Ibid. pp25-27

The reasons which the study team considered may lead decision-makers to select either a higher or lower discount rate than that currently applied in U.S. water resources planning were summarised in Table 4.3 of the Paper (see p70, Appendix B). This table also displays some of the implications of either action.

2.3. Cost Sharing

Paper 5A (WRC, 1975d) presented the results of the study team's work on the development of three cost sharing policy options for implementation and for operation, maintenance and rehabilitation (OM&R) of U.S. Federal and federally assisted water and related land programs and projects. This was the single major task involved in the Section 80(c) study. Paper 5B (WRC, 1975e) presented corresponding policy options for cost sharing of Federal water and related land planning activities.

The major part of the study effort was devoted to describing the existing level of non-Federal cost sharing for implementation and OM&R expenditures (Paper 5A). This description differed from previous cost-sharing studies in two important respects: (i) non-Federal cost sharing rates were expressed in purpose or output categories rather than in terms of measures (dams, levees, channels) within purpose categories, and (ii) the *timing* of non-Federal contributions was taken into account so that alternative arrangements could be expressed in comparable terms (6 percent was used as the interest rate).

Cost sharing in the Section 80(c) data base was defined as *effective composite*. The term 'effective' means that subsidies such as deferred payments were recognised; 'composite' means that both implementation and OM&R expenditures were included.

The Section 80(c) study showed that within the Federal government there are seven departments with 18 agencies and seven independent agencies that have water programs (WRC, 1975 f). The 25 individual agency water programs are funded through a total of 70 appropriation accounts. Because these agencies have different missions, clientele and organisational arrangements, it is not surprising that the study found considerable variation in the non-Federal cost sharing arrangements. These varied widely, by and within purposes, as shown in Table IV.

An important part of each of the three cost sharing options developed by the study was 'the separation of the issue of financing water resource investments from the issue of cost sharing, that is, who ultimately pays for investments. Each of the three options differs from the others in the financing arrangements which it envisions as well as the levels of non-Federal cost sharing' (WRC, 1975 d, p1).

TABLE IV - VARIATION OF U.S. NON-FEDERAL COST SHARING ARRANGEMENTS

IMPLEMENTATION, OM&R AND COMPOSITE COSTS

Sources: WRC 1975d, 1976

Purpose	Average Non-Federal Cost Sharing (%)			Composite Non-Federal Cost Sharing Range (%)
	Implementation	OM&R	Composite+	
Hydropower generation	64	56	64	*
Municipal and industrial water supply	62	78	64	54-100
Water quality (point source)	35	100	64	3-92
Drainage	42	86	46	*
Urban flood damage reduction	17	40	20	13-94
Irrigation	11	92	19	18-53
Rural flood damage reduction	7	34	11	7-80
Navigation	7	8	7	*
All purposes	24	58	30	

+ includes both implementations and OM&R costs

* not given

The three options, developed on the basis of who should finance water resource activities, were:

> '1. The Consistency Option. This option provides for the modification of existing cost sharing rules and practices with minimum changes to attain consistency in the effective cost sharing among all measures and all agencies which produce similar benefits;
>
> 2. The Federal Financing Option. This option provides for Federal financing of all water resources programs and projects with the sale of the vendible benefits to users and with a uniform sharing of residual project costs; and
>
> 3. The Joint Venture Option. This option provides for Federal/non-Federal joint venture financing of all water resources programs and projects with the sale of the vendible benefits to users and with a uniform sharing of both costs and benefits'.
>
> *Ibid. pp71, 72*

A detailed assessment of the effects of each of these options on the financing and cost-sharing arrangements of U.S. water resource investments is presented in Section IV of Part 5A (pp. 69-91) and Section III of Part 5B (pp. 16-23) of the study's report.

3. IMPLEMENTATION OF PRINCIPLES AND STANDARDS AND INTEGRATION WITH EIS PROCEDURES

Since 1973, a considerable amount of work has been undertaken by the U.S. Water Resources Council and various Federal water agencies on the development and implementation of procedures to integrate the multi-objective planning requirements of the Principles and Standards (P & S) and the environmental impact statement procedures of the National Environmental Policy Act of 1969 (NEPA). The work undertaken at the policy level by the U.S. Water Resources Council and at the implementation level by the U.S. Army Corps of Engineers is briefly reviewed in the following sections.

3.1. U.S. Water Resources Council's Study

A Council task force has recently completed a study on the possible integration of certain provisions of the Principles and Standards and NEPA. Although the final study report has not been released as yet, it is understood that the long-term goal recommended by the task force is for complete integration of the EIS requirements of NEPA into the Principles and Standards (WRC, 1975). As an interim goal, the task force has recommended concurrent development of the EIS during the P & S process. In this way, the two processes would 'come together' at certain points in the planning process and greater efficiency would thereby be achieved, e.g., data collection and public review procedures would be undertaken concurrently for both P & S and EIS processes. This task force recommendation is now under active consideration by the Council.

3.2. Corps of Engineers

The experience of the Corps of Engineers in implementing the Principles and Standards and integrating the EIS procedures of NEPA into the overall planning process has been summarised in recent papers by Donovan (1976) and Otto (1977)*. They note that the Corps has translated the broad guidelines of P & S and NEPA into a series of seven integrated engineer regulations '... to enable the conduct of multi-objective planning consistent with the Corps' mission and purposes as enunciated through law, policy, and executive direction' (Donovan, 1976). For shorthand purposes, these regulations are generally referred to as the 200 series' or 'planning framework' regulations. Their purpose, as described by Otto, is as follows:

---

* William Donovan is Chief, Planning Formulation and Evaluation Branch, Civil Works Directorate, U.S. Army Corps of Engineers, Washington D.C., Alexander Otto is a member of the Environmental Quality Section of the Planning Division of the Corps.

Basically, P & S, as exemplified by the Corps '200 series', improves past resource planning guidelines by *systematically* relating all aspects of water and land resources planning to both economic and environmental criteria. Actually, they go further, directing that social well-being and regional development considerations be addressed throughout the planning process. Within the context of the evaluation phase of the planning effort, specific reformulation criteria are provided to Corps planners to enable social well-being (SWB) and regional development (RD) to be considered in the most favourable light, consistent with the fact that they are not to be identified specifically as objectives. *However, life, health, and safety components of the P & S social well-being account may be included as planning objectives in accordance with long-standing tradition and responsibilities incumbent on the Chief of Engineers.* The search for nonstructural solutions to resource management problems is emphasised. Additionally, the regulations direct that relevant beneficial and adverse effects of planning and plans be fully displayed in a system of public information accounts.

The regulations emphasise the equality of the two national objectives ... NED and EQ ... thus necessitating a broad planning approach that encompasses both monetary and non-monetary considerations. Under what we refer to as the P & S - NEPA integrated approach, EQ is not to be considered as an 'add on' to planning; rather, *it is integral* to planning. Further, the integration of all major concerns in the planning process emphasises the use of inter-disciplinary skills in the conduct of that process.

Moreover, the essential approach of the entire '200 series' relates not just to P & S and NEPA but is based as well on the broadened Corps planning experiences acquired in recent years on (1) point wastewater studies, (2) in the expanded urban studies program, and (3) the mandate provided by Section 122 of the RH Act of 1970*. The overall approach is predicated on the philosophy that plans, whatever their number, whether nonstructural or structural, should be publicly acceptable, implementable, and feasible from an economic, social, environmental, and engineering standpoint. Such plans should be harmonious with the institutional milieu that they serve.

---

* River and Harbor and Flood Control Act of 1970.

> This approach recognises that a poor or inadequate planning process ... one that does not concisely define the 'planning objectives', one that does not search out and delineate feasible, implementable alternatives to those objectives, and one that fails to fully and responsibly involve the various publics throughout ... can only aggravate inherently complex economic, social and environmental problems. Implementation of these planning-focussed regulations places substantial confidence where it belongs, namely in the professional judgement and integrity of Corps planners as supported by vigorous, open public involvement.
>
> *Ibid. pp1-2 (emphasis added)*

In support of the '200 series' planning framework, the Corps have recently published (Federal Register 23 February 1977) for public comment a subseries of regulations on environmental considerations. Among other things,

> '... this additional extended guidance provides policy for carrying out thorough environmental resource inventories; defines and articulates approaches for depicting biological, cultural, physiographic and aesthetic features of a study area; provides specific guidance for formulating alternative plans which emphasise environmental quality; provides specific guidance for conducting environmental impact assessment; and, finally provides specific guidance on the evaluation of environmental considerations of alternative plans*. Informally, these latter regulations are referred to as the '400 series'.
>
> *Ibid. p2*

In an attempt to institutionalise environmental considerations into Corps planning, a series of special P & S - EQ training sessions was conducted by the Planning Division of the Corps for personnel in the Corps' field offices, Office of Chief of Engineers (OCE), Board of Engineers for Rivers and Harbours (BERH) and Engineer Institute for Water Resources (IWR) from October 1976 to January 1977. These sessions were intended to provide a clear interpretation of the scope and limits of environmental considerations in Corps planning, thereby extending the content of the ten full P & S training sessions which a multidisciplinary Corps training team from the Planning Division provided to some 500 Corps planners in all field divisions, OCE, BERH and IWR throughout April and May of 1976[+].

---

* The final regulation listed (ER1105-2-450) is an important contribution to the development of detailed procedures for *evaluation* of the environmental effects of alternative plans and for *designation* of the EQ plan. It is reproduced in full in Appendix C of this report.

\+ Commenting on the results of this experience, Otto notes that '... full application of these guidelines by practicing Corps planners is more a matter of time, experience, technique and patience rather than any amount of preaching'.

As an indication of the magnitude of the effort that the Corps is currently undertaking in the implementation of P & S-NEPA guidelines, the following statistics from Otto's paper are worthy of note:

- Since NEPA became effective on January 1, 1970, the Corps has filed over 2,600 Draft and Final EIS's and during the last two fiscal years has spent about $25 and $26 million dollars, respectively, in the preparation, circulation and filing of these statements.

- The Corps now employs over 197 environmental scientists who spend 100% of their time on P & S-NEPA activities and another 378 employees who spend about 25% of their time on these activities. (This does not include the scientists in the Corps research laboratories and staff of universities and professional consultant firms).

What has been the effect of all this expenditure of human and financial resources? According to Otto:

> '... *We estimate that about one in every three active Civil Works projects has been modified because of environmental considerations. These changes range from complete elimination of a project to minor project modifications* ... dams have been resited, elevation of pools adjusted, spoil areas relocated, fish and wildlife resources enhanced and losses mitigated, cultural resources preserved, fish passage structures added, highways relocated, levees set back to avoid channelisation of natural streams, greenbelt floodways established, wetlands created and preserved ... just to name a few of the kinds of modifications that have been made.
>
> Without a doubt, NEPA and the P & S have increased and enhanced public access to our decision-making process. Some of the changes mentioned above have been brought about because of public response to the EIS, but many have occurred because of our own consideration of the environment along with engineering, social and economic factors.
>
> When one considers that it takes fifteen years on the average to plan, design and start construction of a water resources project ... and the NEPA has been in effect for less than half that time and the P & S less than that ... one may well realise the problems and difficulties encountered in trying to apply NEPA and the P & S to

projects conceived and authorised long before their enactment. *Notwithstanding, we believe that water resources projects of the future ... those planned under a full application of the P & S - NEPA integrated approach ... will reflect a better balance between environmental, social and economic concerns.*'

Ibid. p2 (emphasis added)

Notes on Corps Procedures

Some points of note relating to the Corps of Engineers' procedures for planning under the Principles and Standards-NEPA guidelines, as defined in the Corps' '200 series' and '400 series' regulations, are as follows:

i) Terminology. The Corps have adopted a different meaning for the term 'planning objectives' to that used by the U.S. Water Resources Council in the Section 80(c) study (Section 2). In this latter study, the term 'planning objectives' was used synonymously with the term 'objectives' as defined in the Principles and Standards (i.e. National Economic Development and Environmental Quality were treated as 'planning objectives'). In Corps terminology, however, NED and EQ are 'national objectives';

> 'the term *'planning objectives'* refers to water and related land resource management needs that are specific to each study. They are derived from analysis of the needs (opportunities and problems) of the study area that can be addressed to enhance the *national objectives* of P & S'.
>
> U.S. Army Corps of Engineers, ER 1105-2-250, para 6B (emphasis added).

A more detailed explanation of the above definition is given in the notes prepared by the Planning Division of the Corps for the P & S training sessions held for Corps personnel during 1976:

> 'Planning objectives are the national, state and local water and related land resource management needs (opportunities and problems) specific to a given study area that can be addressed to enhance National Economic Development or Environmental Quality (ER 1105-2-200), paragraph 4i).
>
> Planning objectives are a means to bridge the gap between the universality of the two national objectives and the specificity of the problems in a given area. While it is not feasible to plan directly to enhance national economic development by increasing the value of the Nation's output of goods and services and improving national economic efficiency, it is feasible to reduce flood damage on a particular stretch of stream

> and by so doing contribute to the above national objective. Thus it is the purpose of planning objectives to provide sufficient specificity to direct the study in a meaningful manner.
>
> Early in the planning process, objectives are likely to be large in number and general in nature. That is the time when the study team should be taking an expansive approach, considering a range of possibilities, while avoiding detail. As the study progresses, it will be appropriate in succeeding iterations to eliminate objectives that will not or cannot be addressed and to further detail those that will be carried through the study and will eventually provide a basis for evaluation of the plan.
>
> Planning objectives may be as broad and diverse as is the range of water and related land resource concerns in the study area. This is consistent with the breadth of most study authorities. Or they may be as narrow as the critical problem that precipitated the study plus any other concerns that could affect or be affected by the measures to address it or the impacts it causes. For example, if a primary concern is reduction of flood damages, and other concerns relate to water quality and a desire for a free-flowing river; the water quality objective may not have to be addressed, however, the other two concerns are highly related and it would be essential to include them both in order to see the trade-offs involved.
>
> Some appropriate statements of planning objectives are: 'increase open space in X county', 'reduce urban flood damages along Y creek', and 'maintain white-water boating on Z river'. (ER 1105-2-220, paragraph 10, page 8).
>
> Thus, the concept of planning objectives provides for a very flexible framework within which the study team may specify the scope of the objectives, and ultimately the scope of the study. It is particularly important that conflicts, whereby one resource use would preclude another, and constraints be appropriately incorporated into the planning objectives'.

This use of the term 'planning objectives' is similar to the use of the term 'components of the objectives' in the Principles and Standards (and in the Section 80(c) study), viz:

> 'At the outset and throughout the planning process, the responsible planning organisation will consult appropriate Federal, regional, State, and local groups to ascertain *the components of the objectives* that are significantly related

> to the use and management of the resources in the planning setting. *These will be expressed in terms of needs and problems'*.
>
> *Principles p14 (emphasis added)*

ii) Environmental constraints. Corps procedures for plan formulation and evaluation now recognise certain absolute environmental constraints, viz:

> 'Several EQ resources are particularly critical to the EQ objective. These resources involve endangered species, cultural elements, air and water quality, various rivers, seashores, parks, refuges, etc., that are of unique natural, cultural, or aesthetic value. These resources are inextricably tied to the health and quality of man's life and sufficient national attention has been focused on them to require passage of Federal laws to protect and enhance their values. Where these resources occur in a study area, their protection is paramount in the EQ objective. *When any of these resources are adversely affected, the plan or measure responsible must be reformulated to eliminate the adverse contribution. If the adverse impact cannot be eliminated, through plan modification or agreement on mitigation with legally responsible agencies, the plan or measure will be dropped from further consideration, regardless of other beneficial contributions to the NED or EQ objectives. EQ resources of critical national significance are:*
>
> 1. Habitat designated critical for threatened or endangered species by the U.S. Departments of Commerce and Interior;
>
> 2. Sites on, or eligible for nomination to, the National Register of Historic Places;
>
> 3. Air and water quality as defined by State and Federal Standards; and
>
> 4. Environmental, aesthetic, and cultural use areas such as parks, monuments, refuges, rivers, seashores, forests, etc., established by State or Federal law.
>
> U.S. Army Corps of Engineers, ER 1105-2-450, para 8g (emphasis added).

These criteria were used by the Corps during early 1977 in their review for President Carter of projects which had previously been authorised but were not yet constructed*. As a result of this review, several projects were recommended for de—authorisation, either because they violated the above criteria, were considered unsafe or were economically unjustified.

iii) Implementable plans

In exploring the range of alternative plans, there is a conflict between the goal of the Principles and Standards for a full display of alternatives and agency desires to not spend scarce resources on plans which are not 'acceptable' or 'implementable' i.e., public acceptability and implementability have tended to become constraints on the planning process. For example, Corps evaluation procedures (ER 1105-2-250) state that:

> 'Acceptability of a plan is determined by analysing its acceptance by concerned publics. A plan is acceptable if it is, or will likely be, supported by some significant segment of the public. However, during reiterations of the planning tasks, every attempt should be made to eliminate, to the extent possible, unacceptability to any significant segment of the public'.
>
> *Ibid. para 8a*

and:

> '... planning should address both national objectives and avoid developing polar conditions between the two. Because NED and EQ are co-objectives, there should be NED reasonableness in the EQ plan and EQ reasonableness in the NED plan ...'.
>
> U.S. Army Corps of Engineers, ER 1105-2-450, para 12

There thus exists a conflict in agency planning procedures between concern for examination of the full range of trade-offs and the limited planning resources available for such analyses.

iv) Mitigation of adverse environmental effects

Plan modification to mitigate adverse environmental effects is subject to the following considerations in Corps of Engineers' planning procedures (ER 1105-2-450):

---

* See Appendix A, p p 47,48

1. Candidate EQ plans will be modified in an attempt to eliminate *all* significant adverse EQ impacts prior to designation as the EQ plan. Retention of existing EQ resources is paramount. Mitigation is acceptable only to the extent that other EQ resources are not adversely affected.

2. Candidate NED plans will be modified to eliminate adverse EQ impacts only to the extent that incremental NED dollar benefits equal or exceed NED incremental dollar costs to an acceptable degree.

d. When identification of new measures or plan modification does not eliminate all significantly adverse EQ impacts, mitigation measures can be identified and evaluated. Mitigation is compensation for EQ resources or management opportunities lost to society as a result of implementation of a plan. Formulation of mitigation measures will be guided by the following considerations:

1. To the extent possible, loss of EQ resources will be compensated with 'in-kind' resources. Where in-kind resources are not available, closely related resources may be considered. As a minimum, EQ losses will be mitigated only by other EQ resources; e.g., loss of wetlands will not be mitigated by development of golf course.

2. Mitigation of EQ losses must not result in significant degradation or irreversible commitment of other valuable EQ resources.

3. Where it is not possible to mitigate EQ losses with other EQ resources or significant degradation may occur to another EQ resource, a decision must be made to either eliminate the alternative plan or to forego mitigation entirely as a trade-off'.

Ibid. para 13

v) Discounting of environmental effects

In Part 4 of the WRC Section 80(c) study reports (WRC, 1975c), it was strongly argued that it is illogical to discount economic benefits and costs without discounting environmental benefits and costs*. The Corps of Engineers' evaluation procedures are based on the opposite philosophy:

'The question has arisen as to whether environmental and other non-monetary values should be discounted. The answer is no. Such discounting would have no meaning. Often the environmental

---

* See Appendix B, p58

condition in project year 50 is more important than today's conditions. Of course, the timing of impacts is important'.

U.S Army Corps of Engineers, 1976

In accordance with the Principles and Standards, other U.S. Federal water agencies have also adopted the approach of not discounting EQ benefits and costs.

4. CONCLUSIONS

This report has outlined recent developments in the application of the 1973 Principles and Standards in the U.S.A., at both the policy and implementation levels of water planning.

At the policy level, the most important developments have been a series of studies undertaken by the U.S. Water Resources Council. These studies have examined:

i) new options for the planning objectives, discount rate and cost-sharing provisions used in the Principles and Standards (the Section 80(c) study);

ii) integration of the EIS requirements of the National Environmental Policy Act of 1969 (NEPA) into the planning requirements of the Principles and Standards, and;

iii) extension of the Principles and Standards to cover both direct Federal programs and Federally assisted (grant and loans) programs.

The major *policy* issues arising from these studies have been:

(a) Does the multi-objective approach adopted in the Principles and Standards provide the best framework for plan formulation and evaluation or would some other methodology such as an 'integrated objective' approach (Section 2.1.2.) or a 'constrained economic maximisation 'approach (Section 2.1.3) provide more useful information for decision-making?

(b) If the present multi-objective planning approach is preferable to other alternatives, should the objectives used in the Principles and Standards (national economic development and environmental quality) be modified and/or augmented by other objectives, such as regional economic development and life, health, and safety objectives?

(c) The present discount rate used in the Principles and Standards (of the order of 6%) is related to the nominal cost of long-term Federal borrowing. It reflects a policy compromise between the interests of water developers and users (who favour a 'social rate of time preference'* of about 3%) and environmentalists and many economists (who favour use of the 'opportunity cost' of water investment, currently about 10%). What is the appropriate role, and corresponding discount rate policy,

---

* defined in Section 2.2.

for water resource investment? Should a discount rate be applied to environmental benefits and costs, as well as to economic benefits and costs?

(d) Examination of the present non-Federal cost sharing provisions for water projects and programs in the United States has shown that there are wide variations in the present ad hoc arrangements, by and within water planning purposes. What cost-sharing policies should be introduced to ensure a more equitable distribution of project benefits and more efficient use of water?

(e) To expedite the planning process, it has been recognised that there is need to integrate the EIS requirements of the National Environmental Policy Act of 1969 into the planning requirements of the Principles and Standards. Although detailed findings of the U.S. Water Resource Council's study on methods of achieving this integration have not yet been released it is understood that the study has shown that about 90% of the EIS requirements of NEPA are compatible with the Principles and Standards but the remaining 10% have very important legal distinctions. Resolution of these differences will be an important issue in achieving a more complete integration of the two processes.

(f) The major Federal water programs outside the present scope of the Principles and Standards are grant and loan programs directed towards improvement of water quality (primarily those funded by the Environmental Protection Agency). In studying the problems involved in extending the Principles and Standards to cover Federally assisted (grant and loan) programs as well as direct Federal programs, the Water Resources Council's study has found that it is possible and desirable to include these programs within the coverage of the Principles and Standards but that much more attention must be given to strengthening the integration of water quality and water quantity planning considerations.

In addition to these preceding issues, which have arisen primarily from water policy studies, other concerns have arisen from *implementation* of the Principles and Standards by U.S. Federal water agencies. Some issues which are apparent from a review of the procedures which have been developed by the U.S. Army Corps of Engineers to implement the Principles and Standards are as follows:

(g) Confusion exists in the use of some of the basic terminology of multi-objective planning. Whereas the Principles and Standards refer to the *objectives* of national economic development and environmental quality, the Section 80(c) study used the term *planning objectives* and, in their planning procedures, the Corps of Engineers use the term *national objectives* for these broad goals of social welfare. The term *planning objectives* is used by the Corps to describe the needs (opportunities and problems) of the study area that can be addressed to enhance the *national objectives* of national economic development and environmental quality; in the terminology of the Principles and Standards and the Section 80(c) study reports, these needs are referred to as the *components of the objectives*. There is a clear need to end this sort of confusion by the development of precise and agreed terminology.

(h) The Corps procedures have identified a set of absolute environmental constraints on plan formulation and evaluation (see p 35). These constraints recognise that certain environmental resources are of critical national significance and must not be violated by Corps projects. If such environmental constraints are as important as Corps procedures indicate, a strong case must exist for generalising this recognition to all water (and, by implication, other resource) planning by explicitly defining such constraints in the Principles and Standards.

(i) A conflict exists between the concern expressed in the Principles and Standards for an examination of the full range of trade-offs in plan formulation and evaluation and the limited planning resources available to agencies for such analyses. As a consequence, public *acceptability* and *implementability* have tended to become constraints on the planning process.

Resolution of many of these issues is still awaiting the outcome of further studies within the Executive branch of government and the transmission of recommendations from the President to Congress. Although definitive answers cannot therefore be given to these issues at this time, the following tentative conclusions have been drawn by the writer from discussions with staff of the U.S. Water Resources Council and various Federal water agencies:

(a) The multi-objective planning framework used in the Principles and Standards (Section 2.1.1) is widely recognised as being the most comprehensive approach yet developed for the formulation and evaluation of water projects and programs; it is most unlikely that this approach will be replaced by a methodology such as the integrated objective approach (Section 2.1.2.) or that U.S. Federal water resources planning will revert to the constrained economic maximisation approach (Section 2.1.3.) previously used. Important advantages of the multi-objective approach include:

(i) it ensures that a wide range of plans, reflecting different relative emphases on economic, environmental and social goals, are formulated and evaluated;

(ii) it avoids the difficult empirical problem of establishing monetary values for non-monetary effects (as required in an approach such as the integrated objective approach);

(iii) it provides *decision-makers* with information on the economic, environmental and social effects of alternative plans and on the key trade-offs upon which decisions must be made (rather than providing *analysts* with benefit/cost numbers upon which project rankings can be based); and,

(iv) it provides a mechanism whereby the affected public can be involved in (and can contribute to) the plan formulation and evaluation phases of the planning process - it is generally accepted that more meaningful public input can be obtained at the project level of planning rather than at the program level (as required in the integrated objective approach).

(b) It appears certain that the Principles and Standards will become the most important single instrument in U.S. water resources policy formulation and project planning. Two recent developments, both initiated at the request of President Carter, bear witness to this prediction:

(i) The Principles and Standards were used in early 1977 as the basis for reviews by the Corps of Engineers and the Bureau of Reclamation of water projects which had previously been authorised and funded and on which, in many cases, construction had already commenced. As a result of these reviews, the President recommended to Congress that eighteen projects be eliminated and five others be reduced in scale; these projects, approved prior to the introduction of the Principles and Standards in 1973, all failed to meet one or more of the economic, environmental or safety criteria contained in the Principles and Standards. Congress subsequently reinstated all but one of these projects, but the message about future project authorisation is now clear to all.

(ii) As noted in Section 1, the Principles and Standards are now in the process of being revised by the U.S. Water Resources Council as part of a review of U.S. national water policy. It is intended that the revised Principles and Standards will provide the framework for the development of a comprehensive national water policy, firm recommendations on which are scheduled for transmission to the President by December, 1977.

(c) It appears generally agreed that the environmental quality objective in the Principles and Standards will be restated to eliminate some of the confusion that has arisen in interpretation of the existing definition of the Principles and Standards. It is difficult to assess whether regional economic development or the promotion of life, health and safety will be added as new objectives in the Principles and Standards, although the latter is, in effect, already being used as an objective by the Corps of Engineers in their flood control planning. It is also difficult to assess the manner in which demand management and promotion of greater efficiency in water use will be incorporated into the Principles and Standards; as noted in Section 1.3, the current Presidential study has been directed to treat water conservation (i.e., demand management) as an objective, although the manner in which this can best be achieved is unclear at the present time.

(d) With the current emphasis on cost sharing and promotion of greater efficiency in water use, it appears highly unlikely that the discount rate policy will be lowered to reflect the thinking behind the 'social rate of time preference' concept. As water allocation problems become more difficult to solve, the discount rate seems likely to be increased to reflect the 'opportunity cost' of scarce physical, as well as monetary, resources. There appears to be little support at this time for the concept of discounting environmental benefits and costs.

(e) The ultimate goal of *total* integration of the EIS requirements of the National Environmental Policy Act of 1969 with the planning procedures of Principles and Standards will not be completely achieved for some time, owing to certain legal incompatabilities between the two processes. It is expected, however, that *substantial* integration of the two processes will be achieved in the near future through, for example, the use of concurrent data collection and public review procedures.

(f) The increased awareness in the United States of the need to better integrate water quantity and water quality planning will result in extension of the Principles and Standards to include water quality programs which are currently outside the scope of the 1973 Principles and Standards.

(g) *A longer term goal which is gaining wide acceptance is the need to broaden water planning to involve concurrent consideration of land, environmental and energy matters.* It is recognised that the current practice of making decisions along usual program lines needs examination to determine if such decision processes provide adequate solutions to the complex problems of today and the future. The impact of related matters such as land, environment and energy on water management and the development of unforeseen and unwanted consequences among these interdependent elements are among the important reasons why some decisions produced by current institutions are not solving problems in an adequate manner. A start is now being made in the United States in learning how to make decisions involving a larger system of inter-related matters; one important example is the *Water for Energy* program (involving an assessment of the interaction between the nation's water and energy needs and supplies) being undertaken by the U.S. Water Resources Council on behalf of the U.S. Energy Research and Development Administration.

Implications for Australian Applications

The major implications of the preceding review for Australian water planning are:

1. The evolution of the Principles and Standards in the United States has resulted in the promotion of environmental quality as an important concern in water planning programs. The current review of the Principles and Standards will almost certainly result in concern for the efficient use of water being added as an important objective of water planning. Both considerations are of growing importance in Australia and hence the United States experience in integrating these concerns with economic and social objectives should prove of continuing interest to Australian water planners.

2. The practical implementation of the Principles and Standards in the United States has been closely linked with the EIS requirements of the National Environmental Policy Act of 1969. The implication of this for Australian conditions is that whenever State or Federal environmental impact assessment procedures require the preparation of an Environmental Impact Statement, this should be undertaken within a multi-objective planning framework. It is the experience of the Corps of Engineers that, provided the environmental quality analysis is properly undertaken, only

a small amount of additional work is required above that required for the EIS procedure* and the more comprehensive nature of the resulting plan formulation and evaluation procedures helps to ensure that the results will be more readily accepted by the community.

3. As water allocation conflicts become more intense in both the United States and Australia, the need for better integration of water quantity and water quality management programs becomes more pressing - the lower Colorado and the River Murray are excellent examples of such need. The current work being undertaken to strengthen the integration of water quantity and water quality planning considerations in the Principles and Standards should be directly applicable to Australian conditions.

4. The cost-sharing study conducted during 1974/75 by the U.S. Water Resources Council has provided U.S. water planners with a valuable data base on non-Federal cost-sharing arrangements for water resources planning, implementation and OM&R (operation, maintenance and rehabilitation) expenditures, in addition to policy options designed to produce a more equitable distribution of project benefits and more efficient use of water. A similar study designed to establish a data base on existing cost-sharing arrangements between government and private investment in Australian water resource development (e.g., proportion of implementation and OM&R expenditures borne by State governments and users of irrigation water) would provide extremely valuable data for government planning and policy formulation.

5. The planning procedures of the largest U.S. Federal water agency, the Corps of Engineers, have extended the environmental concerns of the Principles and Standards to the establishment of a set of environmental constraints on plan formulation, i.e., the preservation of certain well-defined environmental resources (such as habitat designated critical for threatened or endangered species by the U.S. Departments of Commerce and Interior) is now regarded as an absolute constraint on the planning process. There is need for debate within the Australian community whether a similar set of constraints should apply in Australian water planning and, if so, what should the constraints be and how should they be implemented?

---

* Since the enactment of the National Environmental Policy Act of 1969, U.S. Federal water agencies have been obliged to strengthen the multidisciplinary basis of their planning staffs in order to undertake the preparation of the Environmental Impact Statements required by the Act. The availability of such expertise has thus enabled agencies to implement the Principles and Standards without incurring additional significant cost.

REFERENCES

DONOVAN, W. (1976) New Economic, Social and Environmental Requirements and Related Considerations in Corps Planning. Paper presented at U.S. Army Corps of Engineers Social Scientists Conference, held at Memphis, Tennessee, 20 - 24 September, 1976.

O'BRIEN, W.T., THORNLEY, A.A., & ATKINS, A.S., (1977) Multi-objective Planning for Water Resources Management. Australian Water Resources Council, Project 72/47 June 1977 (Canberra).

OTTO, A.C. (1977) P&S - NEPA Integration : Corps Implementation. Paper presented at Conference on Environmental Impact Analysis, held at University of Illinois, Urbana, Illinois, 22 - 25 May, 1977.

UNITED STATES ARMY CORPS OF ENGINEERS (1975) Engineer Regulations 1105-2-200, 1105-2-210, 1105-2-220, 1105-2-230, 1105-2-240, 1105-2-250, 1105-2-921 *Federal Register* November 10 (Washington, D.C.).

-(1977) Engineer Regulations 1105-2-420, 1105-2-421, 1105-2-440, 1105-2-450 *Federal Register* February 23 (Washington, D.C.).

UNITED STATES WATER RESOURCES COUNCIL (1973) Principles and Standards for Planning Water and Related Land Resources *Federal Register* Vol. 38, No.174 September 10 (Washington, D.C.).

- (1975) Planning and Cost Sharing Policy Options for Water and Related Land Programs (Washington, D. C.)*.

-(1975a) Part 3A - Options for Planning Objectives: Alternatives to the Present Statement (Washington, D.C.).

- (1975b) Part 3B - Options for Planning Objectives : An Integrated Objective Approach to Multiple Objective Planning (Washington, D.C.).

- (1975c) Part 4 - Options for the Discount (Interest) Rate (Washington, D.C.).

- (1975d) Part 5A- Options for Cost Sharing : Implementation and OM&R Cost Sharing for Federal and Federally Assisted Water and Related Land Programs (Washington, D.C.).

---

* Includes individual volumes listed as 1975a - 1975f.

- (1975e) Part 5B - Options for Cost Sharing : Planning Cost Sharing for Federal and Federally Assisted Water and Related Land Programs (Washington, D.C.).

- (1975f) Part 8H - Supplementary Information : Summary of FY1974 Federal Water Resource Obligations by Region, Purpose, Type of Activity, Activity with Type of Financing for each Agency (Washington, D.C.).

- (1976) Testimony by Warren D. Fairchild, Director, before the Appropriations Sub-Committee on Public Works, United States Senate, Fiscal Year 1977 May 26 (Washington, D.C.)

APPENDIX A ADDRESS BY U.S. SECRETARY OF THE INTERIOR ANDRUS TO THE NATIONAL CONFERENCE ON WATER, ST. LOUIS, MISSOURI, MAY 24, 1977

The air of expectancy here today reminds me of the old question so often put to politicians: 'What have you done *for* me lately?'

Only this time the question has a twist: 'What more can you do *to* me now?'

What I hope that I can do to you today is recruit you to help shape and implement a new national water policy in tune with this final quarter of the 20th Century.

For years there has been a great deal of talk about the need for a national water policy, but so far we have made little progress. Today I am going to lay down nine points which I believe will get us moving.

First, I want to discuss with you the recent review of the water projects and explain why I feel the urgent need to launch some initiatives and scuttle some old practices.

Ardent advocates of water projects remain in shock as a result of the President's review of the Bureau of Reclamation and Corps of Engineers projects.

The media called it a 'hit list'.

Emotional responses ranged from alarm and consternation among project proponents to glee and elation among project opponents.

But the decisions made by the President were not based on emotions. He followed a sound course of public policy consideration.

The President in making his decisions, and I in advising him regarding Reclamation projects, did not act out of naivete or lack of understanding of the politics and economics of Federal water resource project authorisation. Both of us as former governors have plenty of experience in this area.

But the President was also aware of the changes taking place in society ... the increased sensitivity to environmental values ... the unwillingness of the general public to accept government decisions on resource use without question.

He knew the public wanted more participation.

He had promised the Nation he would be prudent in public spending, that he would eliminate wasteful use of public resources. Inflation was rapidly escalating costs of water projects, making them candidates for fiscal review, if not prime suspects of having excess 'fat' which could be trimmed from the Federal budget.

Above all, he knew that the electorate had been questioning the old ways. People were asking why decisions made long ago or over a long period of time could not be reconsidered ... why not take a new look under current-day values?

All of the projects reviewed had been under criticism as unsafe, environmentally unsound, or economically unjustifiable. Many of these criticisms had been rejected summarily by project sponsors.

Other arguments were dismissed on 'grandfather' rules ... challenges to projects were considered invalid because they were contrary to rules and acts of Congress prevailing at the time the projects were authorised. In my way of thinking, this is somewhat like ... after choosing our pathway to cross the street ... closing our eyes until we reach the other curb.

Well, I don't plan to operate with my eyes closed. Any decision made at any time by government should be open to scrutiny, and we have to be willing to change our minds and change our course of action when we see a mistake has been made.

When President Carter decided to take a fresh look at the projects, he asked the Office of Management and Budget and the Council on Environmental Quality to provide current economic and environmental criteria. He sought the advice of the Secretary of the Army and myself as to the construction and expenditure status of the projects and the implications of the review.

As a result he sent a message to Congress on February 21 recommending that no funds be provided for the specified projects until they had been reviewed. The President also asked Secretary Alexander and me to carry out an evaluation of projects 'and to develop comprehensive policy reforms in this critical area'. Acting on our findings, the President recommended elimination of 18 projects and cutting back five others.

Our review was thorough even though of necessity conducted on an accelerated time scale. Many of you probably testified at the public hearings. The review teams consisted of experienced staff with access to all records, data and information, and to all projects, supporting calculations and the substantial expertise available within the respective agencies.

My recommendations to the President reflected reasoned and informed judgement based on the individual project reports. I am sure Secretary Alexander feels the same.

The criteria used in the review reflected concepts, standards and values expressed in recent legislation ... most of it enacted since the projects under review had been authorised. There was nothing drastic in the criteria.

The economic criteria took into account current prices on both the benefits and costs sides of the benefit to cost ratio and the current discount rate.

Our environmental standards were in line with recent legislation and today's concern for protection of our quality of life.

Our objective was to determine how well these projects would meet today's needs and the expectations for the future. Obviously, many of the projections of need made years ago when the projects were first formulated no longer are valid.

The water projects review does not constitute a rejection of the past nor a denial of the significant contributions of water resource projects in the development of our Nation and the quality of our lives. Hundreds of projects have provided needed irrigation, drinking water, flood control, electric power, transportation, fish and wildlife enhancement and recreation.

This review was conducted not to eliminate water projects because they have a *low* priority, but because the Administration places a *very high* priority on water policy.

This review marks a renewed awareness and sensitivity toward water resources, and a commitment to examine each water resource issue and to seek its resolution.

It recognises that water is limited in availability just as energy and other resources. But the review declares that the day of considering money to be the only solution to water problems is over. We want results, not in the form of more dams and canals and the like, but in the form of rational use of this precious resource.

Dams, reservoirs and canals do not create water. Since they take years or decades to plan and build, they certainly are not emergency remedies to water-short areas today. Coping with a drought like 1977 requires programs of relief, co-operation, short-term rationing, sacrifice, and management.

The time is upon us when we have to meet the problem head on ... when we have to convince people that water conservation is their only salvation in certain areas ... the government cannot rescue them from either waste or refusal of states and localities to effect land and water use policies which protect the public.

Water is a finite resource, and what the government can do to deliver it where needed is also finite.

It is essential that we keep our perspective and not commit to long-term projects of questionable value because we have an immediate shortage of water in certain areas of the country.

It is clear that policy changes must be made:

... to ascertain that projects under construction, in the planning stage, or proposed in the future will meet an actual need and give broad benefits to the public,

... to assure that we do not waste tax money to benefit a few,

... to ensure that projects are safely designed, constructed and operated,

... to guarantee that the people have meaningful input into the decision-making process,

... to avert unnecessary environmental damage,

... to co-ordinate programs so that we carry out a national policy and do not construct projects which benefit one sector or group while damaging another sector or group.

We have a great opportunity to achieve these changes.

In his environmental message yesterday, President Carter ordered the Office of Management and Budget, the Council on Environmental Quality and the Water Resources Council to review current water resources policy. We will, after consultation with the Congress and the public, report concrete policy recommendations to the President within six months.

I am pleased that the WRC has been assigned a major part in this task. The Council will act as a board of review of agency reports that contain innovative policy reforms or suggestions. The Council will work also with those reports that have major unresolved problems of co-ordination or interagency conflict. The Council will recommend further analysis when necessary, and will reject reports which are inadequate.

At the outset, I promised nine points which I believe should be addressed in achieving a rational and comprehensive water policy.

My first recommendation is that the planning and evaluation criteria expressed in the 1973 *Principles and Standards* of the Council be revised to reflect the President's policy initiatives and concepts. And then, the revised Principles and Standards should provide the framework for the Council's reviews.

Among modifications to the Principles and Standards should be revisions which would:

(1) Add water conservation as an objective for consideration along with the national economic development and environmental quality objectives.

(2) Give equal standing to plans which achieve our environmental conservation and economic objectives through nonstructural means and with explicit consideration of the alternative uses of the waters. Flood plain management and flood insurance should have high priority.

Another change in the *Principles and Standards* is imperative ... abolition of the so-called grandfather clause for all projects not yet underway. This provision now exempts all projects authorised before October 1973 and certain other projects well along in planning and local commitments. All projects not now approved for construction should be subject to the *Principles and Standards.*

I realise that the grandfather provision for the interest rate is a part of the Water Resources Development Act of 1974, and the law would have to be amended to remove it.

A second proposal I make today is that projects that have been authorised for more than eight years and not yet funded or under construction should be de-authorised. This would assure an up-to-date analysis before re-authorisation and would enable us to assure that they meet the test of fitness for the situation as it is known today. Legislation to accomplish this will be submitted to Congress.

De-authorisation is not unprecedented.

You will recall that the Pick-Sloan plan was authorised in 1944, but in 1964 Congress voted to require new authorisation for any projects in the plan which had not yet been initiated.

The large number of projects in the grandfather and the not-yet-started categories indicates that the water resources planning process generates more projects than the economic system and Federal budget can handle.

My third point is that States and other non-Federal entities must be required to share a greater percentage of the burden in financing.

The availability of Federal money provided under very favourable conditions has led to the construction of projects that would not have been built if greater local cost-sharing were required. Present policy provides an irresistible temptation to request projects that would otherwise never be considered. The results too often may be poor planning, unwise and unneeded construction, and inequitable distribution of benefits.

The system of financing Federal projects is sadly out of proportion and out of date. Individual project purposes vary in their requirements for repayment by direct beneficaries from 100 per cent to zero, with the largest Federal subsidy going to such uses as flood control and navigation. While irrigation pays part of its cost, in terms of dollars spent there is a tremendous subsidy. Consequently, there are obvious inequities and the impact on the Federal budget is too large. And the system encourages waste where the price of water is below the actual cost.

A new system of financing must be proposed that will rely on non-Federal sources to pay considerably more of the costs than at present through water rates, capital contributions and other means.

Each project purpose will be affected, for only a comprehensive approach can bring about the balanced results necessary. The data gathering has been done by the WRC task force of two years ago for a study requested by Congress in 1974 legislation. Accomplishment of this goal has been too long delayed.

A fourth need, one of our most desperate, is for the reform of laws, regulations and practices governing allocation of water resources. Our review revealed that these institutional features vary so widely across the Nation that it is difficult to promote efficient use and conservation of water. Reform of these institutions will be difficult.

On a national average, groundwater provides one-fourth of the fresh water. In some states, such as California and Arizona, it provides more than half.

Yet, other than through a tedious process of litigation, in some states there is no system of controls over how or at what rate groundwater is used. Nor is there a management process for inter-relating groundwater and surface supplies.

The Federal government is involved in numerous projects to provide surface supplies to 'rescue' users from groundwater overdraft situations. Often, the new supply is provided below actual cost to induce users to shift from the depleted groundwater system but with no provision for integrated management. Obviously, this unwise resource practice creates more problems than it solves.

Certainly, where the Federal government is muddying up the water, we must revise our programs so that we promote rather than sabotage water conservation.

As for the States, I would hope that they could begin co-operative efforts to bring laws, rules and institutions governing water into the 20th Century. If this does not occur, eventually the Federal government will be encouraged to step in, and another area of State prerogatives will be lost. As a former governor, I don't want to see that happen, and I do not propose any such drastic expansion of Federal influence at this time.

As I indicated a moment ago, the Federal government desperately needs to co-ordinate its programs to encourage wise water use and discourage waste. So my fifth point is the need to scrutinise Federal programs to determine where we can employ these programs to encourage wise water use.

Some ways we can do this is to:

... Require individual water meters for residences in all cities receiving water from Federally-aided systems.

... Set water efficiency requirements for irrigation uses for those receiving Federal assistance in agriculture.

... Establish water efficiency standards for various types of construction where Federal funds are involved or where Federal guarantees are made for the mortgages.

We have a great many tools at all levels of government which can and should be used to promote efficient water use and to discourage waste.

My sixth point is another tough one ... Indian water rights.

In the past, the existence of Federally-secured Indian water rights and their early priority dates have been largely ignored from the standpoint of water resources policy development, planning and use.

Today, the legal entitlement of Indian tribes to the use of water from the systems serving their reservations can no longer be overlooked or discounted. The Nation faces a conflict between the legitimate right of Indians to develop their resources and the impairment of enormous capital investments already made by non-Indians in the same water supply.

Quantification of 'reserved rights' for Indian reservation and other Federal lands is essential to rational planning for the future. The more we continue to plan and use our water resources without regard to determination of Indian and reserved water rights, the more we compound the problem.

My seventh point concerns the need for evaluation of water quality and wastes management planning with conventional water resources allocation and development planning. Waste water reuse, artificial groundwater recharge, minimum streamflows, and safe drinking water are primary objectives ... not just afterthoughts or appendages.

This entire realm must be subject to evaluation through application of the revised Water Resource Council principles and appropriate standards covering all water resources programs.

More Federal funds are being committed now in these programs than in all the conventional water resources development programs combined. The water resources research community should be able to make a large contribution here by technological means, alternative evaluations and management arrangements.

My eighth point concerns safety.

The President has ordered a complete review of all practices related to site selection, design, construction and operations. We also are looking at procedures by which safety problems are identified, analysed and solved.

Reports of these reviews must be completed by October with proposed safety guidelines. The reports and guidelines will then be reviewed by a panel of non-Federal experts who will advise the President by October 1978.

We will not take safety for granted. It must be established and proven in all cases.

And my final point to a large degree is a summary of several other key points ... the primary goal is conservation and more efficient use of water.

In its 1973 report, the National Water Commission noted that 'national water policy has evolved over most of the past century as if water had no cost and there were no limits to its availability'.

Our recent review and the drought should have dispelled this grand illusion.

Water development is costly, and water is certainly not abundant ... at least not everywhere. The existence and construction of large dams and reservoirs does not guarantee abundance ... witness the empty and near empty reservoirs in drought areas.

But we have learned another important lesson from this drought. That is that people can when they try live with considerably less water than they are accustomed to.

I do not suggest that everyone should be cut back to the level of those unfortunate people suffering with a meager ration of water. But their experience does point up the fact that the quantity of water used by most Americans includes a great deal which is in excess of the basic essential supply ... and a great deal of waste.

Just as Americans are learning the facts about energy, they also will have to understand that although it is a renewable resource, water has its physical limitations as to quantity in time and place. We should have the wisdom, and recognise the obligation, to use water carefully.

Water is one of the key resources in the development of America ... it is one of the reasons we grew rapidly into a wealthy and powerful Nation. In the early days of our history, it must have appeared that water like other resources of this continent was inexhaustable.

But water too often has been misallocated, misused and wasted ... here and in other Nations.

Charles Warren, who headed the U.S. delegation to the U.N. Water Conference will be reporting at our luncheon on the results of that conference.

I understand that recommendations were made for action which we will want to consider seriously as we evolve our own national water policies.

The world can no longer enjoy the luxury of waste of any resource.

This morning I have laid out for you the major points I believe are essential to put us on the road toward a comprehensive and realistic national water policy.

What we are talking about in essence is caring ... caring about the environment, caring about wasted tax money, caring about the safety of fellow citizens, caring about the quality of own own lives and the lives of our children.

We have to care enough to change some of our ways ... ways which seemed to have served well in the old days but which no longer are adequate.

There is no better place or time to start this new approach than right here and now.

APPENDIX B 'OPTIONS FOR THE DISCOUNT RATE', SECTION 80(C) STUDY REPORT, PART 4, U.S. WATER RESOURCES COUNCIL, DECEMBER, 1975.

INTRODUCTION

The purpose of this paper is to present options for the interest rate formula (discount rate) used in the formulation and evaluation of Federal water and related land resource projects covered by the WRC Principles and Standards. The information in this paper is to assist the Water Resources Council in recommending an interest rate policy to be used in Federal water and related land resources planning under the Principles and Standards for the next five years.

The interest rate formula currently used in the formulation and evaluation of Federal water and related land resources projects is provided for in the Water Resources Development Act of 1974. Section 80(a) of this Act states that the formula shall be the one approved by the President on May 15, 1962, published as Senate Document 97 on May 29, 1962, as amended by the regulation issued by the Water Resources Council and published in the *Federal Register* on December 24, 1968. This formula provides that the interest rate

> shall be based upon the average yield during the preceding fiscal year on interest-bearing marketable securities of the United States which, at the time the computation is made, have terms of 15 years or more remaining to maturity. Provided, however, that in no event shall the rate be raised or lowered more than one quarter of one percent for any year.

For FY 1976, the formula produces a rate of 6.946 percent. However, as the law provides that the rate may not be changed by more than one quarter of one percent for any year, the applicable FY 1976 rate is 6 1/8 percent or one quarter of one percent more than the FY 1975 rate of 5 7/8 percent.

The discount rate established by this formula is applicable to 'Federal water and related land resources projects' except those exempted by Section 80(b) of P.L. 93-251. The full text of Section 80(b), including this exemption is as follows:

> In the case of any project authorised before January 3, 1969, if the appropriate non-Federal interests have, prior to December 31, 1969, given satisfactory assurances to pay the required non-Federal share of project costs, the discount rate to be used in the computation of benefits and costs for such project shall be the rate in effect immediately prior to

> December 24, 1968, and that rate shall continue to be used for such project until construction has been completed, unless otherwise provided by a statute enacted after the date of enactment of this Act.

The options discussed in this paper provide for a fixed rate of interest to be used for a five-year period rather than one which changes annually. This would eliminate the costly and time-consuming annual recomputations of benefits and costs which are of little practical significance for small changes in the interest rate. Fixing the rate of interest for a five-year period simplifies the planning process and also strengthens the role of the interest rate in the design of water resource projects.

II. THE ROLE OF THE DISCOUNT RATE IN PLANNING

Evaluation of Economic Benefits and Costs

The economic evaluation of water resource projects, planned according to the WRC Principles and Standards, consists of a comparison of the economic benefits and costs of the project. The economic evaluation of projects from a national perspective is reflected in the national economic development (NED) objective of the WRC Principles and Standards. Projects whose benefits exceed their national economic development costs are considered worth-while in terms of the national economic development objective. In a situation where there are alternative projects, the project with the greatest NED benefits in excess of costs (i.e., with the largest net NED benefits), is considered the best project for the NED objective.

To compare benefits and costs, they must be measured at a common point in time. Because project benefits and costs occur at different points in time, they must be adjusted. This adjustment is necessary because receiving a dollar in the future is not worth as much as receiving a dollar today; this is because society prefers income now rather than later. Similarly, paying a dollar of cost in the future is not as burdensome as paying a dollar now.

The factor by which future benefits and costs are adjusted downward so that they can be expressed in comparable terms (present values) is called the discount rate. The term 'discount rate' can be considered to be synonymous with the term 'interest rate'. The higher the rate of interest, the less valuable, in terms of present value, are benefits which would be received in the future. Similarly, the higher the rate of interest, the less costly, in terms of present value, are expenses which do not have to be incurred until sometime in the future.

Characteristics of the Discount Rate

The first characteristic of the discount rate is that it is a 'real' rate. The Principles and Standards require an economic evaluation not a financial evaluation of projects. An economic evaluation is the comparison of the 'real' benefits and cost of a project. That is, the economic benefits in terms of goods and services received are compared to the economic costs in terms of goods and services required to produce the benefits. This is done by measuring the benefits and costs of projects in constant dollars, e.g., 1975 dollars. Because the effects of inflation (a rise in the general price level) are excluded from the estimates of project benefits and costs, each dollar of costs or benefits represents the same purchasing power regardless of when it occurs. Similarly, the discount rate used to express future benefits or costs as present values should be a real discount (interest) rate, that is, a discount rate in terms of goods and services as opposed to money. For this reason, observed market rates of interest - that is, nominal rates of interest - are not directly applicable to water resource projects.

The second characteristic of the discount rate is that it should be a 'risk-free' rate. In evaluating water resource projects, risk and uncertainty must be considered. There are several ways of dealing with risk and uncertainty in plan formulation and evaluation. Although one way is to use a discount rate which includes a risk premium, there are objections to this approach. There are different risks and uncertainties inherent in different water resource projects. Thus, no single discount rate can be selected to account for the specific risks and uncertainties associated with individual water resource projects. Moreover, risk and uncertainty are not necessarily a function of time and, consequently, cannot be appropriately handled through an adjustment in the interest rate. For these reasons, the discount rate should be a risk-free rate, and risk and uncertainty should be dealt with explicitly in the estimation of project benefits and costs. This approach is taken in the WRC Principles and Standards.

Relationship of Discount Rate to Environmental Quality Objective

Economic analysis is only one aspect of multiple objective planning and project evaluation, and it encompasses only benefits and costs that are measured in monetary terms. It is possible, in a mechanical sense, to apply the discount factor to environmental quality and other beneficial and adverse effects which are simply enumerated but not valued in monetary (or any other) terms. For example, at a 10 percent discount rate, 11 rabbits next year would be equal to 10 rabbits only. The results of such calculations have an obvious meaning only when values are attached to the effects, that is, to the change in number of rabbits produced by a plan. If values are attached to environmental effects, then the discount rate would express the relationship between, for example, the value of rabbits in the future and the value of rabbits today. Thus, one could say that 11 rabbits next year *are worth* 10 rabbits today. Because the Principles and Standards call for a display of but not a valuation of EQ and other non-market effects over time, the discount rate is not applied to these effects. The incongruity of using the interest rate for one objective and not another is obvious. And, given the function of the interest rate, it would be logical to apply it to both objectives provided values can be assigned to effects associated with each objective.[1]

The Principles and Standards require formulation of alternative plans which address the NED objective and the EQ objective as well as mixes of the objectives. The purpose of developing these alternative plans is to show the possible trade-offs between the two objectives.

---

1 See, in this connection, U.S. Water Resources Council, Part 3B -- *Planning Objectives: An Option for an Integrated Objective Approach,* November 1975.

(The trade-off is what must be given up for one objective in terms of the other one). Although the discount rate is applied only to the monetary benefits and costs associated with the NED objective, the discount rate policy affects the opportunity cost of obtaining EQ (and other non-market) benefits in terms of net NED benefits foregone for projects formulated under the Principles and Standards.

How the trade-off between NED and EQ is affected by a change in the interest rate is complex and is sensitive to the way in which the costs and benefits of the NED plan are displayed. When the net benefits of the NED plan are expressed as annual net benefits (which is the normal agency practice) as opposed to present values, different effects of changing the discount rate on the trade-off can be obtained.[2] However, in general, raising the discount rate reduces the opportunity cost of obtaining EQ benefits in terms of net NED benefits foregone regardless of whether a present value or average annual equivalent approach is used; that is, according to the Principles and Standards, raising the discount rate makes it easier to justify environmental quality enhancement.

Formulation of Water Resource Plans

There generally are alternative ways of utilising resources. Each alternative will produce different types and patterns of benefits and costs over time. Different discount rates will change the economic evaluation of what constitutes the best use of water and related land resources. In this way, the discount rate affects the nature and direction of that part of the water resources program covered by the WRC Principles and Standards.

In theory, the discount rate influences the formulation and economic evaluation of water resource projects planned under the Principles and Standards in five specific ways:

> First, at the broadest level, the discount rate affects the economic evaluation and, thus, the resulting decision of whether to develop water resource projects for economic purposes as opposed to a decision to do nothing.
>
> Second, the discount rate affects the evaluation of the optimal size or scale of projects which develop water resources for economic purposes.
>
> Third, the discount rate affects the optimal mix or set of economic outputs provided by projects because different purposes also differ in their time pattern of benefits and costs.

---

2 'Annual net benefits' is defined as the constant annual value which, when discounted, equals the present value of a project's net benefits. For a 100-year period of analysis, the equivalent annual value of a present value of $100 is $10 at a 10 percent discount rate; it is $3.20 at a three percent rate.

Fourth, the discount rate affects which kind of alternative measure can best serve a given purpose (e.g., a channel versus a reservoir for flood control).

Fifth, the discount rate affects the optimal combination of resources employed in the construction and maintenance of measures (e.g., concrete versus earthen channels).

It should be noted that for water resource projects which are not planned under the Principles and Standards, the discount rate affects only two of the above. These are the combination of resources used in the construction and maintenance of measures (number five) and the kind of measure which best serves a given purpose (number four). For example, in the EPA waste treatment construction grant program, the level of treatment is determined by Federal laws and regulations. In this case, the role of economic analysis is to identify the least cost way of achieving the specified requirement.[3] That is, where the outputs of a plan are specified, the problem is one of cost-effectiveness analysis rather than of benefit-cost analysis.

The total Federal water resources program is a combination of (1) agency programs which are subject to the Principles and Standards and (2) agency programs which are not subject to the Principles and Standards. Because of this, the evaluation of effects on the total water resource investment program resulting from changes in the interest rate is highly uncertain. Therefore, discussion of the programmatic impact of changes in the discount rate in Section IV will focus on those agency programs covered by the WRC Principles and Standards.

Conceptual Framework for Choosing a Discount Rate Role of Government Investment

The basic aim in the use of government investment funds for water resources development should be to maximise the contribution of these investments to social welfare or, in words of the Principles and Standards, 'the quality of life'.

Government should select worth-while investments which are thought to be inappropriate for the private sector to undertake. Private investment may not be forthcoming or sufficient where external social benefits are either too technically difficult or too expensive to capture. In addition, worth-while projects which are exceptionally large or risky may not be attempted by private business. As an agency of society, it is felt that government should nurture and support the desired activities and goals of private citizens and private business; government should not merely pursue its own profit-making ends. For the majority of Federal programs, the decisions made in attempting to achieve this goal are based largely on the collective judgements of the people as represented by Congress and the President.

---

3. The capacity of the plant is also subject to economic analysis. In providing a reserve for future growth, the choice of a discount rate affects the trade-off between adding an increment to capacity now and the present value of providing the capacity at the end of the design horizon.

In the early years of water resource planning and development, the major explicit concern appeared to be the economic welfare of society. Often, opportunities to improve the non-economic aspects of social welfare were either ignored, buried in a project whose economic returns were high enough to cover the cost of non-economic extras, or evaluated in monetary terms through estimated shadow or proxy prices. Ultimately, the explicit concern for these non-market outputs, which also contribute to the quality of life, led to the enactment of the WRC Principles and Standards. National economic development and environmental quality have become equally important objectives. Under the Principles and Standards, only the alternative plan formulated to serve the national economic development objective must have net economic benefits. A recommended plan need not have net economic benefits if the deficit in net benefits is the result of serving the environmental quality objective.

Concepts of the Interest Rate

There are different theories regarding the appropriate discount rate for use in water resources planning. Moreover, there are market interest rates which could be referenced in selecting a rate for use in water resource planning. The purpose of this section is to provide a conceptual discussion of the interest rate in the overall context of a market economy.

In a perfectly functioning, risk-free economy, a single market rate of interest would result from the interaction of the demand for investment funds and the supply of funds from saving. The demand for investment funds reflects the productivity of physical investments while the supply of funds from saving reflects individual time preferences regarding current consumption of goods and services versus future consumption. The equilibrium market rate of interest in this simple case would reflect both the marginal rate of time preference and the marginal productivity of investment. In other words, the return earned by the last dollar of consumption foregone would equal the return from the last dollar of investment.

In an imperfectly functioning market economy, adjustments must be made to both the demand for investment funds and the supply of funds from saving in order to identify the socially optimum interest rate for public investment decisions. These adjustments are required because the private market does not take into account all of the social consequences of investment decisions. First, the social productivity of investment is the private productivity of investment adjusted up or down for external effects. Unfortunately, there are no empirical measures of the generalised external effects of private investment. Second, the net yields to investors may be less than the private productivity of capital because of corporate income taxation. Third, risk and uncertainty reduce the demand for investment funds. Fourth, private time preference may not reflect collective wishes to provide for the future.

Since the extent of the needed adjustments is unknown, analysts have emphasised certain concepts for actually selecting a discount rate for government investment decisions. While there are many different arguments, they can be grouped into three broad categories: the social rate of time preference, the opportunity cost of investment and the cost of Federal borrowing.

Those who are concerned that the amount of both private and government investment made annually by society is too 'small' often refer to the concept of social time preference. The income and well-being of future generations may be increased if we were to increase our investments in productive capacity now. It is held by some that the private time preferences which affect market rates of interest do not reflect sufficiently our collective wishes to provide for future generations. If we knew everyone else was willing to sacrifice more to provide for future generations, then we might also be willing to do so. Other analysts argue that the private sector is myopic and does not invest enough for the future. If we accept these arguments, market rates of interest would be considered too high for use in government investment decisions.

Another view of social time preference focuses on specific project outputs rather than the level of total investment. According to this view, the discount rate represents a choice of weights regarding the relative value of specific public project benefits over time. Thus, the social rate of time preference may be higher as well as lower than market rates of interest[4].

Those who are concerned with the productivity of water resource investments relative to the alternative uses of resources emphasise the notion of opportunity cost. These analysts emphasise that government investments in water resources utilise resources which could alternatively be employed elsewhere for other government investments, for private investments or for personal consumption. In this view, water resource investments are not considered to be unique in any essential way. Thus, the return on water resource investments should be as great as the return from opportunities which are foregone by such alternatives. In certain cases, water resource projects displace specific private investment opportunities. For example, government development of a hydro-power site may pre-empt the use of that site for private development. In other cases, no specific private investment is foreclosed but, nevertheless, the resources required to construct and operate the project have an opportunity cost because they could be employed elsewhere.

Within the opportunity cost approach, there are many variations. For example, some analysts focus on the return on private investments as the proper measure of opportunity cost; others also include private consumption along with private investment in defining opportunity cost.

A third concept is the cost of Federal borrowing. This is the basis for the current interest rate policy. It must be recognised that one inherent difficulty in the current approach is that the WRC discount rate, which should be a real rate, is set by a nominal (market) rate. The rationale for this concept is that the Federal Government should earn at least as much on its investments as it costs the government to borrow funds. If the government invested in all of the water resource projects which have a return at least equal to its cost of borrowing, the

---

4 Some advocates of this second approach to social time preference would account for the opportunity cost of resources withdrawn from the private sector by using a 'shadow price' for these resources.

government would maximise its net benefits, i.e., its 'net worth' in water resources. This would not necessarily maximise the net worth of society if the water resource investments precluded other investments with a greater productivity.

Use of Empirical Data

The opportunity cost of water resource projects in terms of alternative investments in the private sector, the rate of private (as opposed to social) time preference, as well as the cost of government borrowing are reflected in empirically observable data. Such data are displayed in Appendix A.

These data are not directly usable in selecting an appropriate interest rate for formulating and evaluating water resources projects for two reasons: returns on private investment as well as market interest rates are in nominal as opposed to real terms, and they also reflect the risk and uncertainty inherent in investment decisions in the private sector. Because the appropriate discount rate for water resources planning is a real, risk-free rate, adjustments to empirical data must be made to account for risk and inflation.

Another problem in referencing empirical data is the choice of an appropriate time period for empirical analysis. Because the rate of inflation and interest rates have changed dramatically in the past few years, different periods of analysis give different results.

It also should be pointed out that the current interest rate formula looks at the nominal yields on very long-term government securities during the previous fiscal year. Changing either the time period or the classes of securities included in the calculation would lead to a different value for the nominal cost of borrowing for the Federal Government.[5]

III. OPTIONS FOR THE DISCOUNT RATE

This paper examines the impacts of three levels of the discount rate on the formulation and evaluation of water resource projects within the programs covered by the WRC Principles and Standards. In addition, the alternative interest rates can be related to but do not uniquely follow from the concepts of social time preference, opportunity cost and the cost of Federal borrowing. These analyses should aid decision-makers in selecting the appropriate discount rate for use in water resource planning for the next five years. The underlying assumption of the impact assessment approach is that the shape, nature and magnitude of Federal investment in water resources is primarily a policy determination, not simply a technical decision. The rates to be assessed are three, six and ten percent.

---

5. As of June 30, 1975, the outstanding amount of the securities used in the current interest rate formula was $16.8 billion. This is about 5 percent of the outstanding amount of marketable government securities which was $315.6 billion.

The three percent rate corresponds roughly to the rate used by agencies during the early years under Senate Document 97. It approximates the rate on U.S. Government bonds and high grade corporate bonds during the 1940's and 1950's (see Appendix A). A three percent rate would satisfy most of those who believe that a low interest rate is appropriate for economic evaluation of Federal water resource projects. The justification for use of low interest rates is usually based upon a belief that the rate of interest used in water resource investment decisions should be related to the social rate of time preference and that the social rate of time preference is less than private market rates of interest.

The six percent rate is close to the prescribed WRC rates of 5 7/8 percent for FY 1975 and 6 1/8 percent for FY 1976, which were determined by means of the formula provided for in Section 80(a) of the Water Resources Development Act of 1974 (P.L. 93-251).[6] As discussed in the Introduction, the current WRC rate is related to the nominal cost as opposed to the real cost of long-term Federal borrowing. It also provides a point of reference to the existing situation from which this study can depart in its consideration of options for the discount rate.

The 10 percent rate approximates the level desired by many of those who believe it represents the rate of return on investments in the private sector.[7] It would satisfy many of those who have expressed the need for a high rate to insure that government water resources investments generate a yield comparable to the real returns on investments in the private sector.

IV. ASSESSMENT OF IMPACTS OF THE OPTIONS

Only a qualitative assessment of the impacts on water resource projects and programs from changes in the discount rate can be made in this paper. Very little is known, at present, about the sensitivity of the economic evaluation of water resource projects to large changes in the interest rate when reformulation of projects at the different interest rate is permitted. In reformulation, the scale of the project as well as the mix of outputs and the measures used to achieve the outputs of the plan can be changed. While the discount rate is conceptually an important determinant of project design, other factors, such as the mix of objectives being addressed, legislative authority, cost sharing policies, engineering standards and agency policies, may have a much greater effect on project design. Empirically, little is known about the relative importance of each of these. However, to the extent that other considerations constrain the ability of planners to reformulate projects according to economic criteria, raising the interest rate reduces the net economic benefits of projects more than the amount they would be reduced if projects were fully reformulated according to economic criteria.

---

6 The formula prescribed by Section 80(a) of P.L. 93-251 actually yields a rate of 6.946 percent for FY 1976.

7 See U.S. Water Resources Council, *Proposed Principles and Standards for Planning Water and Related Land Resources*, and *Federal Register*, Volume 36, No.245, pages 24166-24167, December 21, 1971. The Council concluded that average real rate of return on marginal physical investments in the non-Federal sector is around 10 percent. Also see Budget Circular A-94 Revised, 27 March 1972, subject: Discount Rates to be Used in Evaluating Time Distributed Costs and Benefits, which states that 'The prescribed discount rate of 10 percent represents an estimate of the average rate of return on private investment, before taxes and after inflation'.

Because the Principles and Standards set forth both national economic development and environmental quality as objectives, we would expect economic analysis to play different roles in plan formulation as the mix of objectives that are to be served varies. While it should be a consideration in the development of plans to serve the EQ objective as well as plans to serve mixes of the two objectives, it should certainly dominate all other considerations in the formulation of the NED plan which is the plan which maximises net economic benefits.

Water resource projects, like many other types of investments, are typically characterised by large initial installation costs relative to their annual operations and maintenance costs. The expected benefits of projects seldom are projected to be constant over time. A typical pattern is for benefits measured in constant dollars to increase over the life of the project. In Table 4-1, Part A focuses on the effect of discounting benefit streams which begin at various points in time.

Raising the discount rate reduces the value today of any given stream of benefits. While raising the discount rate affects every benefit stream, it has the greatest relative effect on benefit streams which begin further out in the future as opposed to those which begin sooner. In addition, an increase in the discount rate will reduce the present value of benefits more than costs because of the difference in their timing. By reducing the present value of project benefits more than the present value of project costs, an increase in the interest rate reduces the number and scale of projects which can be economically justified for implementation.

In Table 4-1, Part B illustrates some of the impacts, from the standpoint of individual projects and overall programs, of alternative discount rates. Project scale is affected because increments to projects with benefits realised only in the future become uneconomic, and the more distant in time the benefits are relative to project costs, the more sensitive is the economic feasibility of projects and project increments to a change in the interest rate.

Table 4-2 shows the effect of the discount rate on the value today (discounted or present value) of measures which have different relationships between annual OM&R costs and first costs. Water resource projects constructed in the past typically had particularly high first costs relative to their annual operations and maintenance costs. This is in part a reflection of the interest rate used in plan formulation. However, the relationship between first costs and annual costs is not usually the same for all measures which can be used to solve a given water resource problem. Thus, when the interest rate is raised, measures whose annual O&M costs are high relative to first costs can be substituted for those measures whose costs do not have this relationship. This can be brought about by a change in the kind of measure used and by changes in the mix of resources used to construct and maintain the measure even though the expenditures over time for the measure may be greater. For example, at higher rates of interest, not only might a channel be substituted for a reservoir, but earth may be substituted for concrete in the channel design.

In each planning situation, each of the potential measures has outputs with an expected stream of benefits over time. Each of the possible measures which can provide an output also has a different time stream of costs. As a result, the combination of outputs with the largest

TABLE 4-1

IMPLICATIONS OF DISCOUNTING PROJECT BENEFITS

PART A: VALUE TODAY OF BENEFITS AMOUNTING TO $1.00 PER YEAR BEGINNING IN VARIOUS YEARS AND ENDING IN THE 100TH YEAR

Beginning Year of Benefits	Three Percent Discount Rate	Six Percent Discount Rate	Ten Percent Discount Rate
Today	$32	$17	$10
10th Year	23	9	4
20th Year	17	5	1
35th Year	10	2	0
50th Year	6	1	0

PART B:

ILLUSTRATIVE PROJECT AND PROGRAM IMPACTS a/

Category of Impact	Three Percent Discount Rate	Six Percent Discount Rate	Ten Percent Discount Rate
Projects	Leads to larger projects. Justifies, for example, an investment of as much as $17 today to receive an annual benefit of $1 (for 80 years) beginning as late as 20 years from today.	Justifies, for example, an investment of $5 today to receive an annual benefit of $1 (for 80 years) beginning as late as 20 years from today.	Leads to smaller projects. Justifies, for example, an investment of only about $1 today to receive an annual benefit of $1 (for 80 years) beginning as late as 20 years from today.
Programs -- Assume that the dollar cost of justified projects is greater than available funds.	Tend to build larger NED-oriented projects but also fewer projects.		Tend to build smaller NED-oriented projects but also more projects.
--- Assume that the dollar cost of justified projects is less than available funds	Leads to larger number of NED-oriented water resource projects and higher program level for water resources. Reduced funds made available for other uses (current consumption or alternative private or public investments).		Leads to smaller number of NED-oriented water resource projects and lower program level for water resources. More funds made available for other uses (current consumption or alternative private or public investments).

---

a/ Focuses on project scale impacts alone; other impacts are discussed subsequently.

TABLE 4-2

IMPLICATIONS OF DISCOUNTING PROJECT COSTS

PART A:

COST TODAY (DISCOUNTED COST) OF $100 OF EXPENDITURES OCCURRING OVER 100 YEARS OF PROJECT LIFE 1/

Ratio of One Year's OM&R Cost to First Cost	Three Percent Discount Rate	Six Percent Discount Rate	Ten Percent Discount Rate
Zero Percent	$100	$100	$100
1/2 Percent	77	72	70
2 Percent	54	44	40
5 Percent	43	30	24
10 Percent	37	24	18
100 Percent	32	17	10

1/ Assumes first costs occur in first year and annual OM&R costs are evenly distributed over the remaining 99 years.

PART B:

ILLUSTRATIVE PROJECT AND PROGRAM IMPACTS

Category of Impact	
Projects	Discounting project costs leads to a preference for measures whose costs are incurred over time as opposed to today. As the discount rate is increased, the relative attractiveness of measures with deferred costs as opposed to measures whose costs are incurred today is increased.
Programs	Raising the discount rate shifts costs of water resource programs to future generations because at higher rates of interest, the relative attractiveness of measures with high annual OM&R costs is increased.

net economic benefits (i.e., the plan which combines the measures that best serve the NED objective) will change as the interest rate changes.

In summary, a change in the interest rate affects the nature of the water resource program covered by the WRC Principles and Standards. This change is felt most strongly in projects oriented toward the NED objective. A change in the interest rate causes different kinds and different numbers of NED-oriented projects to emerge from the planning process. These differences include different outputs, different measures to provide the outputs, different project sizes, as well as different patterns of costs and benefits over time.

Raising the discount rate favors smaller NED-oriented projects with high initial benefits (e.g., current water supply versus future water supply, reduction of flood damages to existing development versus future development). Moreover, a high discount rate favours producing outputs with measures with high operation and maintenance costs and low capital or investment costs. Conversely, lowering the discount rate favours larger NED-oriented projects whose benefits may not be realised until the future (e.g., development of a new waterway which would require several years for traffic to develop). Moreover, a lower discount rate favours measures with high initial investment costs and low operation and maintenance costs.

Although these conclusions focus on NED-oriented projects, a change in interest rate policy can also be expected to impact on projects formulated for the EQ objective or for a mix of NED and EQ objectives. For example, an increase in the discount rate should lead to the substitution of O&M costs for capital costs in all kinds of projects, EQ-oriented as well as NED-oriented. Paradoxically, however, an increase in the discount rate may result in larger EQ-oriented projects as well as a greater tendency to recommend these projects in some cases because the Principles and Standards do not require discounting of EQ benefits. This happens because a rise in the interest rate will often reduce the costs of obtaining EQ enhancements in terms of NED benefits foregone.

If the construction budget for programs covered by the WRC Principles and Standards expands or contracts in accordance with the number of justified projects, a set of conclusions can be drawn. As the interest rate rises, the total investment in projects oriented toward NED would decline, and the proportion oriented towards EQ would rise. There would be fewer projects as well as smaller projects oriented toward NED. Moreover, projects of all kinds tend to have higher OM&R costs. In this case, the size of the investment program would likely be constrained by the interest rate.

If the constraint on the water resource program is, instead, the budget level (that is, there are more justified projects at 'any' investment rate than can be funded), another set of conclusions can be drawn. The use of a high interest rate would permit more NED-oriented projects to be constructed because NED-oriented projects would tend to be smaller, and each dollar of investment expenditure would tend to place larger OM&R costs on future generations (which also enjoy the benefits). Conversely, the use of a low interest rate would permit fewer projects to be constructed because NED-oriented projects would tend to be larger. Moreover, each dollar of investment expenditure would place smaller OM&R costs on future generations.

V. 'PREFERENCE SETS' FOR DISCOUNT RATE OPTIONS

The effects discussed in Section IV, although important to water resource projects and program formulation, do not provide a complete basis for choosing an interest rate. Interest rate policy decisions should also consider the broader implications of the decision in the context of the interest rate in the allocation of national resources. From that point of view, choosing an interest rate which is lower than the opportunity cost of capital represents a loss to the Nation in that investments with inadequate returns are undertaken and national economic growth depressed. At the same time, choosing an interest rate which is higher than the social rate of time preference represents a loss to the Nation in that desirable public investments would be precluded.

Table 4-3 summarises the reasons which may lead decision-makers to select either a higher or lower discount rate than that currently applied in water resources planning. This table also displays some of the implications of either action.

The first entry in each preference set addresses the issue of time preference. Since the issue of social time preference can be resolved only on the basis of a subjective response by decision-makers, the implications of each option in terms of time preference are shown through illustrative examples.

The preference set which would lead to a decision to lower the discount rate shows that choosing a lower rate would result in less difficulty in justifying NED-oriented projects within programs covered by the Principles and Standards. There would be more situations in which NED-oriented projects could be justified and NED-oriented projects would tend to be larger. (It has also been pointed out that a lower discount rate permits modification of a project to accommodate non-NED concerns while still maintaining net economic benefits for the project as a whole).

Because a low discount rate would increase the number of economically justified NED-oriented projects, it would provide a large number of projects from which to select projects for construction within current or increased funding levels.

The cost of using a lower discount rate is, of course, that the rate of return on NED related water resource investments would decline. Moreover, projects for either the NED, the EQ or any other objective would be more capital intensive, and, thus, fewer projects would tend to be built at each possible level of implementation funding.

The preference set which would lead to a decision to raise the discount rate shows that selecting a higher rate would place greater emphasis on EQ-oriented concerns and projects within programs covered by the Principles and Standards. The greater emphasis on EQ results because EQ benefits and costs are not discounted under the Principles and Standards. Thus, in general, raising the discount rate would reduce the opportunity costs of obtaining EQ benefits, i.e., costs in terms of NED benefits foregone. Furthermore, a high discount rate favours EQ to the extent that a higher rate reduces the scale of or eliminates NED-oriented projects which would result in environmental damage.

TABLE 4-3

'PREFERENCE SETS' FOR DISCOUNT RATE OPTIONS

Preference Set I

In Order to Lower the Discount Rate to Three Percent, Decision-Makers Should Want to:

1. Invest a dollar today to receive an annual one dollar benefit for 16 years beginning as late as 84 years from now[1], or alternatively invest as much as 32 dollars today to obtain an annual benefit of one dollar for 100 years;
2. Make it less difficult to justify NED-oriented projects within programs covered by the WRC Principles and Standards;
3. At the program level, provide a large number of NED-oriented projects from which to select projects for construction funding; or
4. Increase the implementation budget for agencies covered by the WRC Principles and Standards to permit construction of more projects.

And be Willing to:

1. Accept a lower real rate of return on water resource investments for the NED objectives;
2. Accept a large backlog of authorised but unconstructed projects unless program funding levels are increased; and
3. Accept larger NED-oriented projects and more capital intensive projects for either the NED or the EQ objective.

Preference Set II

In Order to Retain a Discount Rate of Six Percent, Decision-Makers Should Want to:

1. Invest a dollar today to receive an annual one dollar benefit for 53 years beginning as late as 47 years from now[1], or alternatively invest as much as 17 dollars today to obtain an annual benefit of one dollar for 100 years.

And be Willing to:

1. Accept status quo with regard to discount rate.

Preference Set III

In Order to Raise the Discount Rate to Ten Percent, Decision-Makers Should want to:

1. Invest a dollar today to receive an annual one dollar benefit for 76 years beginning as late as 24 years from now[1], or alternatively invest as much as 10 dollars today to obtain an annual benefit of one dollar for 100 years;
2. Place greater emphasis on EQ-oriented concerns and projects within programs covered by the WRC Principles and Standards;
3. At the program level, require a higher real rate of return on water resource investments for the NED objective and reduce the number of NED-oriented projects which can be selected for implementation; and
4. Perhaps decrease the implementation budget for agencies covered by the WRC Principles and Standards.

And be Willing to:

1. Forego NED-oriented opportunities to develop water and related land resources which cannot be justified at higher rates of interest;
2. Accept restudy of unauthorised projects formulated under lower rates of interest. Authorised projects may also need to be restudied depending upon decision made regarding "grandfather" clause; and
3. Accept smaller NED-oriented projects and less capital intensive projects for either the NED or the EQ objective.

---

1 Assumes a 100-year period of analysis; for a 50-year period of analysis at three percent, the annual benefit could begin as late as the 46th year and last four years; at six percent, the annual benefit could begin as late as the 37th year and last 13 years; the annual benefit could begin as late as the 23rd year and last 27 years.

If EQ benefits and costs were valued in monetary terms like NED benefits, a higher discount rate would not encourage EQ investments particularly since EQ benefits often are deferred into the future. If all project effects were valued and discounted, the use of a high discount rate would lead to the selection of only high return water and related land resource investments for all objectives.

The use of a high discount rate results in a higher real rate of return in investments made for the NED objective. Because many of the projects formulated at lower rates of interest would no longer be economically justified (i.e., have a rate of return at least equal to the discount rate) at a higher rate of interest, these projects would have to be restudied if the new interest rate policy were to be applied to them. Finally, a higher interest rate may reduce the number and cost of the projects justified under the Principles and Standards. This effect could reduce the implementation budget for agencies covered by the Principles and Standards although the more probable effect would be a reduction in the backlog of unfunded projects.

The alternatives of 3 and 10 percent discussed in this paper bracket the range of choice for an interest rate policy. However, within this range there are considerable differences of opinion regarding the appropriate rate. These differences reflect different views of the conceptual basis for interest rate policy as well as of the role of water resource investments.

In establishing an interest rate policy, it is important to note that the interest rate is only one determinant of the character and direction of the Federal and federally-assisted water and related land resources investment programs. Very little of the federally-assisted investment program is currently subject to the Principles and Standards and, thus, need not be economically justified. Moreover, in the direct Federal investment program (which is subject to the Principles and Standards), the interest rate is explicitly considered only on the cost side in the formulation of EQ-oriented plans, and these plans need not have net economic benefits to be recommended under the Principles and Standards. Consequently, interest rate policy is not a particularly useful policy variable for controlling program levels. Therefore, determination of the interest rate policy should focus on effective use of whatever funds are allocated to water resource investments. The role of the interest rate in guiding the formulation of individual projects is an important one in assuring that scarce resources are effectively utilised. However, it should be kept in mind that, according to the Principles and Standards, the interest rate performs this function with regard to project effects associated with the NED objective but, as explained earlier, the same result is not achieved in project effects associated with the EQ objective.

APPENDIX A

TIME SERIES: INTEREST RATES, RELATION OF PROFITS AFTER TAXES TO STOCKHOLDERS EQUITY, AND RATE OF INFLATION

Year	Water Resource Plan Formulation (Fiscal Year)	US Government Bond Yields--Long Term (Calendar Year)	Short Term (3-5 Yrs) US Government Security Yields (Calendar Yr)	Total Corporate Bond Yields (Calendar Yr)	Relation of Profits After Taxes to Stockholders Equity (Percent, Calendar Yr)	Rate of Inflation
1975	5.875	--	--	--	--	--
1974	5.625	6.99	7.81	8.98	14.5	9.7
1973	5.5	6.30	6.92	7.80	12.8	3.8
1972	5.375	5.63	5.85	7.63	10.6	2.4
1971	5.125	5.74	5.77	7.94	9.7	4.7
1970	4.875	6.59	7.37	8.51	9.3	5.8
1969	3.250/4.625	6.10	6.85	7.36	11.5	4.7
1968	3.250	5.25	5.59	6.51	12.1	3.3
1967	3.125	4.85	5.07	5.82	11.7	3.3
1966	3.125	4.66	5.16	5.34	13.4	2.5
1965	3.125	4.21	4.22	4.64	13.0	1.7
1964	3.000	4.15	4.06	4.50	11.6	1.1
1963	2.875	4.00	3.72	4.50	10.3	0.4
1962	2.625	3.95	3.57	4.62	9.8	0.7
1961	2.625	3.90	3.60	4.66	8.9	0.5
1960	2.500	4.01	3.99	4.73	9.2	0.7
1959	2.500	4.07	4.33	4.65	10.4	2.2
1958	2.500	3.43	2.90	4.16	8.6	2.1
1957	2.500	3.47	3.62	4.21	10.9	6.0
1956		3.08	3.12	3.57	12.3	6.6
1955		2.84	2.50	3.25	12.6	2.2
1954		2.55	1.82	3.16	9.9	1.0
1953		2.94	2.56	3.43	10.5	1.7
1952		2.68	2.13	3.19	10.3	2.7
1951		2.57	1.93	3.08	12.1	8.1
1950		2.32	1.50	2.86	15.4	2.2
1949		2.31	1.43	2.96	11.6	3.0
1948		2.44	1.62	3.08	16.0	9.6
1947		2.25	1.32	2.86	15.6	14.6
1946		2.19	1.16	2.74		10.4
1945		2.37	1.18	2.87		0.0
1944		2.48	1.33	3.05		2.2
1943		2.47	1.34	3.16		4.4
1942		2.46	1.46	3.34		11.9
1941		2.05	.73	3.34		6.8
1940		2.21	.50	3.55		3.4
1939		2.36	.59	3.77		-1.5
1938		2.56	.83	4.19		1.3
1937		2.68	1.40	3.94		9.0
1936		2.65	1.11	3.87		-0.8

APPENDIX C 'ENVIRONMENTAL CONSIDERATIONS : EVALUATION OF ALTERNATIVE PLANS', U.S. ARMY CORPS OF ENGINEERS. (REGULATION NO. 1105-2-450), FEBRUARY 1977.

1. Purpose. This regulation provides environmental guidance on the evaluation of alternative plans as an extension of ER 1105-2-250.

2. Applicability. This regulation is applicable to all OCE elements and all field operating agencies having Civil Works responsibilities.

3. References.

a. Public Law 91-190, *National Environmental Policy Act of 1969* (83 Stat. 852).
b. Section 122, Public Law 91-611, *River and Harbour and Flood Control Act of 1970* (84 Stat. 1818).
c. Water Resources Council, *Principles and Standards for Planning Water Related Land Resources (P&S)*, 38 FR 24778-24869, 10 September 1973.
d. ER 1105-2-200.
e. ER 1105-2-220.
f. ER 1105-2-230.
g. ER 1105-2-240.
h. ER 1105-2-250.
i. ER 1105-2-507.
j. ER 1105-2-921.

4. General. Evaluation of alternative plans is a vital link in the planning process. It provides both planner and decision maker with specific information concerning the effectiveness of a plan in addressing the water and related land resource problems. It also provides important information on the public's perception and acceptance of the planning effort. Environmental considerations in plan evaluation are particularly crucial to the success of the entire planning effort. Both NEPA and the P&S mandate that full environmental consideration will be employed in the planning process to enhance, preserve, and protect the quality of the environment to the fullest extent possible. This regulation provides guidelines to assure that environmental considerations in the formulation of alternative plans, or lack thereof, are surfaced, evaluated, and where missing, incorporated. The evaluation task will assure input of environmental considerations during early stages of the planning process, redirecting plan formulation to incorporate additional environmental input, when needed.

5. Period of Analysis. The period of analysis used for Environmental Quality (EQ) evaluation can be independent of the period used for other aspects of plan evaluation (ER 1105-2-250). A plan's significant contributions to the EQ account, either beneficial or adverse, may extend or occur beyond the 50- or 100-year period used for the evaluation of other contributions. This is especially critical in evaluating irreversible commitments of EQ resources. Specification of the EQ period of analysis, if different from that being used for the other evaluation effort, will be made in the System of Accounts.

6. The Planning Team. The evaluation of alternative plans will be conducted by a multi-disciplinary planning team. The environmental specialist(s) on the team will participate fully in all plan evaluation efforts to assure that environmental concerns are given full and equal consideration.

7. EQ Considerations in Evaluation of Planning Objective Fulfillment. Overall guidance for determining the relationship of project induced changes to planning objectives is provided in ER 1105-2-250. Planning objectives are water and related land resource management needs (opportunities and problems), specific to the study area, that can be addressed to attain the national EQ and/or National Economic Development (NED) objectives. Planning objectives direct plan formulation in solving water and related land resource problems. To assure full EQ consideration in evaluation of plans which address the planning objectives, the following specific guidelines are given:

a. Significant environmental impacts (from Impact Assessment) are compared to the planning objectives to determine any relationships. To identify relationships and evaluate degree of contribution, the planning objectives should be translatable into specific non-monetary units, such as miligrams per liter oxygen, hectares of marsh, or miles of scenic river. Consequently, a planning objective to improve the oxygen level (in mg/l) of stream X can be related to the change in oxygen levels (mg/l) provided by removing excessive Biochemical Oxygen Demand (BOD) with a wastewater treatment system. If the plan significantly increases the oxygen level in stream X over the 'without' condition and maintains this contribution over the period of analysis, a high degree of planning objective fulfillment has been attained.

b. Significant environmental changes with a plan are compared to projected 'without' conditions for the same resources in impact assessment. This will identify long-term relationships between the alternative plans and future environmental conditions. This relationship will show whether or not specific environmental resource management needs expressed in the planning objectives can be fulfilled by the plan. For example, a planning objective is to improve oxygen levels over the existing condition in stream X, a highly polluted stream. Plan A will initially increase oxygen levels 4 mg/l over the existing condition, a significant change. However, non-point sources will continue degrading oxygen conditions until, by year 2000, the plan's influence on oxygen levels will only maintain the existing level, which was already highly polluted and below State standards. Planning objective fulfillment has been only partially obtained, even though the 'without' condition was improved.

c. The full range of a plan's significant environmental changes will be utilised in the evaluation of planning objective fulfillment. Planning objectives will address environmental needs specific to the study area but attainment of the needed resource management must contribute positively to the EQ objective.

d. The public will be informed of relationships between the study's planning objectives and the national EQ objective, as defined in Principles and Standards (P&S). Relationships between a plan's significant environmental impacts and planning objective fulfillment will be presented and explained to the public for their concurrence or to resolve any disagreement or misunderstanding.

8. EQ Considerations in Evaluating Contributions to the System of Accounts. Overall guidance for evaluating a plan's contributions to the P&S System of Accounts (SA) is provided in ER 1105-2-250. Specific guidelines for preparing the SA are provided in ER 1105-2-921.

a. Impact Analysis. Significant impacts are analysed to decide if they are beneficial or adverse. The full range of significant impacts will be displayed in the pertinent accounts. The impacts are compared to determine the best plan(s) and to identify contributions to the national objectives and other accounts. The EQ and Social Well Being (SWB) contributions are measured in non-monetary and usually dissimilar units, making comparisons difficult. For evaluating contributions to the EQ account, it may be helpful to group similar or closely related impacts (from a measure common to all plans being compared) and identify a 'common denominator' shared by all the impacts. This is generally only useful for two or three impacts that are critical to the decision makers. If the impacts, for example all result in loss of fish, this 'common denominator' could be used to simultaneously compare effects of a number of impacts caused by similar measures.

b. Break-point Analysis. It may be desirable to employ a 'break-point' analysis to contain evaluation efforts within reasonable and productive limits. Break-point analysis can also be used to redirect plan formulation more efficiently in the reiteration process. Essentially, break-point analysis identifies sudden changes in 'value' of additional units of an output in relation to additional 'costs' incurred in producing the added output. For the EQ account, the intrinsic 'value' of increased units of an EQ contribution can be compared to the 'costs' in resources committed or management opportunities foregone to produce the contribution. Where there is a *significant change* in the relationship between the 'incremental value' and the 'incremental cost', it may be possible that increasing the EQ contribution beyond this point is no longer justifiable or desirable. For example, reducing BOD in stream X by 10 mg/l will require converting 2000 hectares of woodlands to grassland for land treatment of wastewater. Further reduction of BOD by another 5 mg/l requires only 100 more hectares of woodland. However, overall reduction of BOD by more than 15 mg/l will require an additional 500 hectares of woodlands for each additional mg/l of BOD reduction. A high public value was placed on BOD reduction by 13 mg/l to meet State standards. Achieving an additional output of BOD reduction by 2 mg/l 'costs' little more in additional land commitment. However, to increase the EQ contribution by reducing BOD a total of 17 mg/l would be very 'costly' in commitment of woodlands without a correspondingly high value associated with the increased output. The break-point at 15 mg/l could direct plan formulation to cease output of BOD reduction at 15 mg/l and avoid the additional loss of woodlands, even though it would be possible to continue increasing positive contributions to one component of the EQ account. The evaluation effort would also be reduced.

c. EQ in the NED Account. Careful examination will be made of any relationships between environmental impacts and economic contributions to the NED account. For example, diminution of a fish run may result in economic losses (loss of fisherman days). This loss must be quantified and counted as an adverse NED impact. The environmental specialist(s) will work closely with the economic specialist(s) in verifying monetary contributions attributed to environmental resources.

d. EQ in the EQ Account. Detailed comparison will be made between significant environmental impacts and their effects on components of the EQ objective. Particularly critical are those contributions that relate to dynamic physical, chemical, and ecological processes maintaining vital life support systems for man and other important biological communities. The role of processes such as seasonal flooding, tides, nutrient cycling, weathering, or biological succession must be carefully evaluated in light of significant project impacts. Natural air and water purification systems are other life support elements that must be looked at closely. Additionally, the availability and quality of aesthetic and cultural resources is, among others, essential determinants of our quality of life. The influence of significant environmental impacts on these resources are, therefore, a very important evaluation consideration. The environmental specialist(s) on the Planning Team must assure full consideration of all aesthetic and cultural aspects in this evaluation effort. Similarly, all contributions of environmental impacts that could irreversible commit any of the EQ resources underlying components of the EQ objective must be fully identified and accounted for in the EQ account.

e. EQ in the Regional Development (RD) Account. As stated in ER 1105-2-921, the contribution of impacts to the EQ account are 'regionalised' in Table 2 of the SA. This eliminates the need to account for regional environmental impacts in the RD account. Where possible, and relevant to the decision on plan selection, contributions of environmental impacts to other RD components must be determined. For example, environmental quality may have influence of regional employment, business, and industrial activity. Similarly, environmental quality has a direct influence on the value of regional recreation; e.g., where fish and game populations, water quality, aesthetics, etc., are of high quality, associated recreational development is worth more. The environmental specialist(s) on the planning team will again assure that full and accurate representation is given to contributions of environmental impacts in the RD account; of course, the extent of such analysis will be limited by the state of the art. Close co-ordination with the economics specialist will be necessary.

f. EQ in the SWB Account. Environmental contributions to the SWB account will relate principally to public health, education, and recreation. Environmental impacts that affect numbers and kinds of disease vectors or nuisance organisms will be carefully evaluated in this account where possible and relevant to the decision on plan choice. Similarly, impacts affecting air and water quality can have very significant influences on human health and well-being. Educational opportunity is also an important component of the SWB account. Natural areas of high quality are in great demand for scientific and educational uses. When this type of area is affected by an alternative plan, the impact's contribution to the SWB account must be fully evaluated. Environmental quality also affects recreational quality and opportunity. When impacts contribute to the quality of the environment, e.g., aesthetics, biological populations, air and water quality, etc., then they can also be greatly influencing recreational opportunity. Detailed analysis must be made of any environmental contributions to this component of the SWB account.

g. EQ Resources. Several EQ resources are particularly critical to the EQ objective. These resources involve endangered species, cultural elements, air and water quality, various rivers, seashores, parks, refuges, etc., that are of unique natural, cultural, or aesthetic value. These resources are inextricably tied to the health and quality of man's life and sufficient national attention has been focused on them to require passage of Federal laws to protect and enhance their values. Where these resources occur in a study area, their protection is paramount in the EQ objective. When any of these resources are adversely affected, the plan or measure responsible must be reformulated to eliminate the adverse contribution. If the adverse impact cannot be eliminated, through plan mofidication or agreement on mitigation with legally responsible agencies, the plan or measure will be dropped from further consideration, regardless of other beneficial contributions to the NED or EQ objectives. EQ resources of critical national significance are:

(1) Habitat designated critical for threatened or endangered species by the US Departments of Commerce and Interior;

(2) Sites on, or eligible for nomination to, the National Register of Historic Places;

(3) Air and water quality as defined by State and Federal Standards; and

(4) Environmental, aesthetic, and cultural use areas such as parks, monuments, refuges, rivers, seashores, forests, etc., established by State or Federal law.

9. EQ Considerations in Applying Specified Criteria. Overall guidance for applying specified evaluation criteria is presented in ER 1105-2-250. The specified evaluation criteria further test the value and responsiveness of alternative plans in solving water and related land resource problems. When plans are not responsive to the criteria, reformulation will be necessary to resolve the conflict. To assure full EQ consideration in applying the specified evaluation criteria, the following guidelines are given:

a. Acceptability. The acceptance of a plan by concerned publics must be determined. Where a significant segment of the public believes strongly that the plan will adversely affect components of the EQ objective, then reiteration will specifically address and attempt to resolve adverse environmental aspects while maintaining the basic orientation of the plan. Determination of acceptability will be based on public input.

b. Completeness. A 'complete' plan incorporates all actions necessary to implement the plan. Where plans enhance environmental quality, comparison will be made between investments and actions needed to accomplish the enhancement and those that are actually in the plan and could be implemented should the plan be recommended and authorised. Where enhancement of environmental quality will require extensive investments or actions by other agencies, or where future actions must be separately authorised, the plan is essentially 'incomplete'. Reiteration shall address specific inclusion of the needed actions, including detailed obligations of others.

c. Effectiveness. The technical performance of a plan in attaining planning objectives will be evaluated in light of 'state of the art' techniques for sound environmental planning in water resources development. Measures that enhance the EQ objective will reflect up-to-date thinking of the professional environmental sector.

d. Efficiency. A plan's efficiency is measured by its ability to achieve planning objectives in the least-cost way. Costs are generally considered in terms of dollars. For EQ purposes, however, 'costs' can also be non-monetary commitments of resources that are lost to society. Irreversible and irretrievable commitments of EQ resources will be considered as environmental costs and will relate to the plan's efficiency. Lost resource management opportunities are also 'costs' that affect a plan's efficiency.

e. Certainty. The certainty of a plan is the likelihood that planning objectives will be achieved should the plan be carried out. Identifying certainty involves two different criteria: (1) determining that there is enough information and data on hand to be able to predict impacts and attainment of planning objectives, and (2) determining if the project is sufficiently operational to attain the planning objectives. Early in the planning process, careful analysis must be made of available data and significant data gaps. An analysis must be made between data needed to identify potential EQ impacts and the data available. It may be necessary to collect additional data before further planning efforts are continued. This evaluation can save considerable and expensive planning efforts from being misallocated. The operational certainty of a plan achieving planning objectives must be evaluated. Where alternative plans contain separable EQ enhancement measures, not inextricably related to traditional project purposes, an analysis will be made of the institutional and implementation requirements necessary to actually carry out these measures. Evaluation will be made of the plan's non-Federal cost sharing requirements and the history of non-Federal support for participating in environmental enhancement programs in the study area. Where considerable non-Federal participation will be required to enhance the EQ objective, the overall certainty of attaining related planning objectives may be affected. This determination will be fully addressed in the evaluation of alternative plans.

f. Geographic Scope. The geographic area to be studied for impact assessment should be large enough to encompass actual, significant environmental (and other) impacts. In some cases, e.g., inland navigation, the area to be studied may be very large, because impacts are widespread. The geographic area for problem identification should be large enough to encompass a reasonable set of alternative plans, from which a final plan will be selected. This area will generally be smaller than the area for impact assessment.

g. NED B/C Ratio. Economic benefits in relation to economic costs are determined. NED elements that are affected by the quality of the environment (man-days of hunting or fishing, commercial fishing, water-oriented recreation, etc.,) will be included in the NED B/C evaluation. Input from the US Fish and Wildlife Service, National Marine Fisheries Service, Bureau of Outdoor Recreation, and other agencies with responsibility in this area will be sought early in the planning process and will be incorporated in the evaluation.

h. Reversibility. The degree to which a given alternative plan, after full or partial implementation, could be reversed and the impacted areas restored to approximate the 'without' condition is determined. Where impacts irreversibly degrade components of the national EQ objective, the plan will be considered irreversible. Particular attention must be given to *cumulative* effects of environmental impacts. While a single impact may not be considered irreversible, the cumulative effect of a number of similar impacts from this and other projects could push a resource past a critical threshold or reversibility.

i. Stability. A plan's stability represents the range of alternative futures that can be accommodated by the plan's output of services. The degree of 'certainty' in obtaining environmental goals in the alternative futures and the degree of 'uncertainty' (para 21, ER 1105-2-921) related to the plan's environmental impacts on each alternative future relate strongly to the stability of the plan.

10. EQ Considerations in Trade-Off Analysis. Basic guidance for incorporating trade-off analysis in the plan evaluation task is given in ER 1105-2-250. In trade-off analysis, the contributions of each plan are compared so that what is gained or foregone by choosing a given alternative plan over other alternatives is clearly set forth. Public involvement in trade-off analysis is crucial. Decisions must be made that could effect the day-to-day lives of study area residents. These decisions must be made based on the active participation and input by local citizens. Study personnel must, however, evaluate the national interest in all trade-off decisions. To assure that full consideration of environmental quality is incorporated in the trade-off analysis, the following guidelines are given:

a. All significant contributions to components of the national EQ objective will be displayed and presented to the public. Where possible, EQ impacts will be shown in numerical terms to emphasise the magnitude of the contribution and to facilitate comparison.

b. In addition to displaying all significant EQ contributions, the public must be shown which environmental resource management needs or opportunities are being gained or foregone by choosing a given alternative. A useful tool for displaying these EQ gains or losses is the 'preference set'. Preference sets can be valuable vehicles for fully involving the public in the decision-making process, allowing them to contribute to the critical trade-off decisions concerning their community's environmental quality. While the preference set is a public participation tool, it can be useful in displaying trade-off analysis in the study report especially for localised projects such as flood control. One possible preference set sequence follows:

(1) Selection of plan X indicates that residents of the study area prefer the following levels of EQ attainment:

(a) Specify degree of management, protection, enhancement, or creation of areas of natural beauty and human enjoyment;

(b) Specify degree of management, preservation, or enhancement of valuable cultural, physical, or ecological resources or systems;

(c) Specify degree of enhancement of land, air, and water quality;

(d) Specify degree of enhancing other EQ components not listed above.

(2) By accepting the above levels of EQ attainment, study area residents must be willing to forego the following resource management opportunities: specify environmental, economic, and social resource management opportunities that will be foregone by achieving levels specified in (1) above.

(3) To achieve the above levels of EQ, study area residents must be willing to make the following commitments:

(a) Specify actual EQ resource commitments that will be required;

(b) Specify significantly adverse EQ impacts that are key determinants in the trade-off decisions:

(c) Specify actual economic commitments and costs that will be required;

(d) Specify actual institutional changes that will be necessary;

(e) Specify expected social conflicts in the study area as a result of EQ attainment;

(4) In light of the above commitments and assumptions, is the alternative plan, as presently proposed, acceptable to the concerned public?

(5) If the alternative plan, as presently proposed, is not acceptable, what changes should be made?

c. Public input to the trade-off analysis will surface alternative plans that are most environmentally acceptable to the local citizens. It will also identify needed changes in planning objectives or plan formulation.

11. Designating the EQ Plan. Overall guidelines for designating the EQ Plan are given in ER 1105-2-230 and ER 1105-2-250. An EQ Plan is one that addresses a study's planning objectives in a way that best emphasises aesthetic, ecological, and cultural contributions. National environmental policy mandates that, among other things, important cultural and natural aspects of our national heritage are preserved and a diverse and varied environment is maintained. Consequently, designation of an alternative plan as the EQ plan implies that the plan best enhances and protects the aesthetic, cultural, and natural diversity of the environment. The minimum requirement for designating a plan from the final array of alternative plans as the EQ Plan is that it must make a net positive contribution to the EQ account when compared to the 'without' condition. EQ Plans must also be viable, implementable, meet evaluation criteria, and not violate planning constraints.

a. An EQ Plan will be designated on the basis of the following priorities:

(1) Designate the alternative plan which provides maximum positive contributions to one or more components of the EQ account and makes no significant adverse contribution to any component of the EQ account which is not mitigated in kind.

(2) Where it is shown that the above cannot be done, then designate as the EQ Plan the alternative plan which provides the greatest net positive contributions to the EQ account. This plan can include mitigation measures and EQ losses can be offset by EQ gains.

(3) In some planning studies it may not be possible to designate an EQ Plan given the minimum requirement in paragraph 11. If and when an EQ Plan cannot be designated, the attempt to do so will be clearly supported and documented. In such cases, the plan that most nearly meets the minimum requirement will be identified and labelled as the 'EQ Oriented NED Plan'.

b. Designation of an alternative plan as an EQ Plan will be subject to the following criteria or constraints:

(1) The 'no action' alternative will *not* be designated as the EQ Plan. A 'no action' alternative provides no more positive contribution to the EQ account than does the 'without' condition.

(2) *No* alternative plan will be designated as an EQ Plan that contains measures which cause adverse modification of a natural, free-flowing stream or river having significant environmental value.

(3) *No* alternative will be designated as an EQ Plan that contains measures that degrade wetlands having significant environmental value.

(4) The planner may find it necessary to place upper limits on the level of EQ enhancement attained by an EQ Plan. An analysis should be conducted to determine a candidate EQ Plan's 'NED reasonableness'. This analysis is to identify any significant 'break-points' where NED costs in producing the EQ output increase significantly without significant increases in the EQ output. For example, there will be a point when the cost of an incremental increase in the amount of dissolved oxygen in a stream becomes extremely costly; or the cost of an incremental increase in the Ph of an acid stream becomes extremely costly.

12. EQ Considerations in Designating the NED Plan. Overall guidelines for designating the NED Plan are given in ER 1105-2-230 and ER 1105-2-250. Since the P&S requires equal national significance be given to environmental quality and national economic development, planning should address both national objectives and avoid developing polar conditions between the two. Because EQ and NED are co-objectives, there should be NED reasonableness in the EQ Plan and EQ reasonableness in the NED Plan. EQ reasonableness in designation of an alternative as the 'NED Plan' will be based upon the following considerations and constraints. (see para 8g):

a. *No* alternative designated the NED Plan will cause the following adverse environmental impacts:

(1) Destruction or degradation of critical habitat for threatened or endangered species, as designated by the US Departments of Commerce and Interior;

(2) Destruction or degradation of archaeological or historical sites on, or considered eligible for nomination to, the National Register of Historic Places without a signed Memorandum of Agreement with the Advisory Council on Historic Preservation, the National Park Service, and the State Historical Preservation Officer;

(3) Degradation of air or water quality below State or Federal Standards, or below existing conditions if already less than these Standards.

(4) Destruction or degradation of environmental quality in any Federal, State, or local governmentally established environmental use area (wilderness areas, refuges, parks, rivers, scientific-educational areas, etc.,) without full agreement between all affected agencies.